Large Camel Farming

Bernard Faye • Gaukhar Konuspayeva •
Cécile Magnan

Large Camel Farming

A Care-Management Guide from Breeding to Camel Products

Springer

Éditions Quæ

Bernard Faye
UMR SELMET, CIRAD-ES, Campus International de Baillarguet
Montpellier, France

Gaukhar Konuspayeva
Biology and Biotechnology Faculty, Department of Biotechnology
Al-Farabi Kazakh National University
Almaty, Kazakhstan

Cécile Magnan
Clinique Vétérinaire équine de Cambajo
Sauzet, France

ISBN 978-94-024-2236-8 ISBN 978-94-024-2237-5 (eBook)
https://doi.org/10.1007/978-94-024-2237-5

Jointly published with Éditions Quæ, Versailles, France
The translation was done with the help of artificial intelligence (machine translation by the service DeepL.com). A subsequent human revision was done primarily in terms of content.
Translation from the French language edition: "L'ÉLEVAGE DES GRANDS CAMÉLIDÉS" by Bernard Joannes Faye et al., © Éditions Quæ 2022. Published by Éditions Quæ. All Rights Reserved.

This Springer imprint is published by the registered company Springer Nature B.V.
The registered company address is: Van Godewijckstraat 30, 3311 GX Dordrecht, The Netherlands

Foreword

It has become trivial to say that camel farming has undergone a major transformation in recent decades, which is not unrelated to the dual challenge of climate change and economic globalization. This is reflected in a remarkable evolution of breeding methods and significant changes in the management of this animal, which are not without effects on its physiology and health. Coming out of its marginality, camel breeding and its products are experiencing a real craze that goes beyond the borders of the arid zones to which it seemed to be eternally devoted.

Although the services of the camel have probably been provided by desert people for thousands of years, it is only recently that products such as milk and wool have entered the international economy, that production systems have intensified, and that the processing of these products (milk, meat, wool, hides) has been modernized. More and more people are willing to embark on the adventure of camel production outside the usual breeding areas to provide products and services that meet the standards of twenty-first-century consumers. There are also more veterinarians who are confronted with the health management and care of an animal that has suddenly been subjected to unusual forms of breeding, and which is little studied in the schools and universities responsible for training the actors of animal health.

This book aims to give some guidance, if not to succeed in the sharing of the experience of camel breeding with full knowledge of the facts and to support all the actors of a fast-growing sector, including veterinarians. This publication does not claim to be exhaustive, but it hopes to be useful to breeders who wish to market products that are rightly or wrongly reputed to be beneficial, as well as to veterinarians and decision-makers who see the camel, once confined to a few marginal regions of the world, now ready to be part of agro-ecosystems that are less restrictive than its original desert.

Photo credits: All photos are by Bernard Joannes Faye, unless otherwise stated in the captions.

Cover photo: Left: Camel calving at Kharj camel farm, Saudi Arabia (credit B. Faye). Center: Camels in Jaco farm, France (credit B. Faye). Right: Milking of camel at Makhanov Otemis farm, Kazakhstan (credit B. Faye)

International Camel Expert, Emeritus expert CIRAD
Montpellier, France
Bernard Faye

Professor, Al-Farabi Kazakh National University
Almaty, Kazakhstan
Gaukhar Konuspayeva

Veterinary Practitioner
Sauzet, France
Cecile Magnan

Introduction (Preface)

The large camelids, which include two domestic species (the dromedary and the Bactrian camel), have been long confined to the arid zones of the Old World, but are increasingly present in arid countries worldwide. Western countries are among these, with a growing presence in the European agricultural landscape. Intended for various tourist activities (cultural events, camel rides), sports (racing), or production (milk, even wool), these species do not belong to a well-established breeding tradition in temperate latitudes. Thus, breeders, technicians, and veterinarians are often confronted with the most basic questions on how to behave toward the animal, how to handle it, how to determine its feeding needs, how to manage its reproduction, or simply how to provide it with the necessary daily care.

It is therefore the goal of the authors to provide useful, simple, and practical answers to not only all members of the traditional sector, fully realized in arid countries, but also to regions where the camel is new. The authors put forth that this is due not only to the remarkable resistance of the large camelids to extreme ecological conditions in the context of climate change, but also to the exceptional quality of its products (milk, meat, wool) and its services (transport, riding, management of pastoral areas). This enthusiasm is also due to the good valuation of its products and services on the national and international markets, to such an extent that we can speak of the emergence of a new sector for a species that until recently was confined to the subsistence of nomadic populations living in the most hostile regions of the Old World, with no significant impact on the local economy.

This book aims to provide a basic knowledge of the physiology of this animal as well as practical advice on how to manage its breeding and health.

This guide will therefore deal successively with (i) general information on the species, (ii) the physiological bases, (iii) the technical management of a camel farm (reproduction, feeding, health), (iv) the economic management of a farm, and (v) its main productions and services.

Montpellier, France — Bernard Faye
Almaty, Kazakhstan — Gaukhar Konuspayeva
Sauzet, France — Cecile Magnan

Acknowledgments

This book was designed within the framework of the CAMELMILK project, which is part of the PRIMA program funded by the European Union (Horizon 2020 research and innovation program) whose partners are Spain, Algeria, Turkey, and France (for the breeding and scientific support part), and Italy, Croatia, and Germany (for the scientific support part only). The CAMELMILK project aims to promote the production and marketing of camel milk around the Mediterranean basin by strengthening the actors of the camel milk sector.

Financial support also came from the CIRAD UMR SELMET within the framework of the European project "Roles of camel breeding in Modern Saharan societies: contributing to their adaptive capacities face to global changes – CAMED." The aim of this project is to describe, understand, and model the past and recent trajectories of camel societies and to strengthen the resilience of camel breeding systems according to pastoral and animal resource management methods and marketing strategies. It includes partners from Algeria, Morocco, and France.

The authors thank Doug Baum from Texas Camel Corps, USA, for his reviewing of the English version of the book.

Camel Milk

CAMELSHIELD
MULTIFONCTIONNALITÉ DE L'ÉLEVAGE CAMELIN
DANS LES SOCIÉTÉS SAHARIENNES

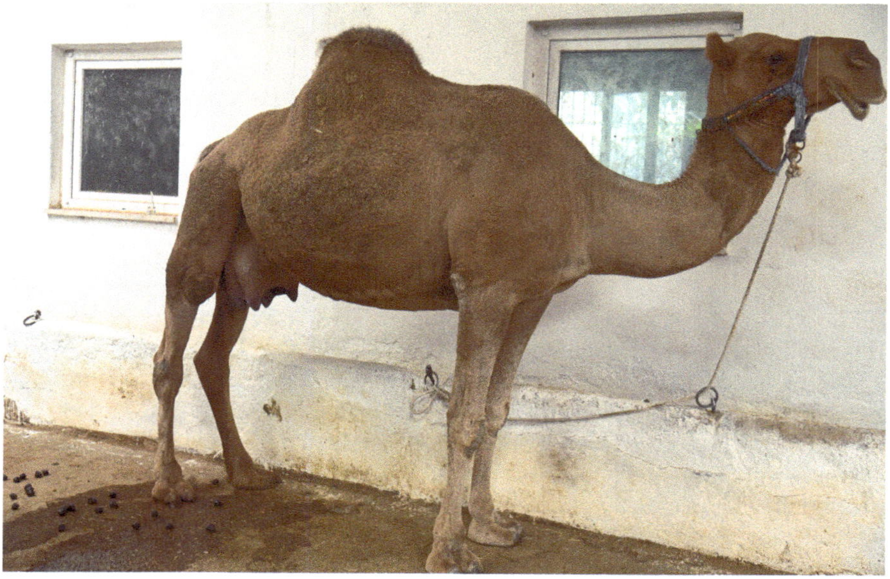

Contents

1 General Information on Large Camelids in the World: Origin, Taxonomy, Biodiversity, and New Settlements 1
1.1 Origin, Taxonomy, and Present Status of Large Camelids in the World 1
1.2 Biodiversity and New Settlements 6
References 12

2 Anatomical Features of Large Camelids 13
2.1 General Compliance 14
2.2 The Skeleton 15
2.3 Dentition 16
2.4 Internal Anatomy 18
2.4.1 Muscular Apparatus 18
2.4.2 Nervous System 19
2.4.3 Lymphoid System 19
2.4.4 Blood System 20
2.4.5 Respiratory System 22
2.4.6 Digestive System 22
2.4.7 Urinary System 25
2.4.8 Locomotor System 26
2.4.9 Abdominal Topography 28
References 29

3 Physiology of Large Camelids: Life Cycle, Adaption to Ecosystems, and Reproduction 31
3.1 The Life Cycle of Large Camelids 31
3.2 Mechanisms of Adaptation 34
3.2.1 Adaptation to Heat 34
3.2.2 Adaptation to Drought 36
3.2.3 Adaptation to Malnutrition 39
3.3 The Physiological Basis of Reproduction 40
3.3.1 The Sexual Cycle in Camels 40
3.3.2 The Sexual Cycle in Males 41
3.3.3 Mating 45

3.3.4 Gestation . . . 47
3.3.5 Parturition (Box 3.1) . . . 48
References . . . 53

4 The Physiological Basis of Large Camelid Lactation, Digestion, and Nutrition . . . 55
4.1 Physiology of Lactation . . . 55
4.1.1 Udder Morphology . . . 56
4.1.2 Milk Flow . . . 56
4.1.3 The Lactation Curve . . . 59
4.2 The Physiological Basis of Digestion and Nutrition . . . 60
4.2.1 Digestion in Large Camelids . . . 62
4.2.2 Nutrition of Large Camelids . . . 63
4.2.3 Feeding Behavior . . . 64
4.2.4 Quantities Ingested . . . 65
4.2.5 Storage and Destocking of Fat Reserves . . . 65
4.2.6 Water Requirements . . . 66
References . . . 68

5 Technical Management of a Camel Farm with a Focus on Dairy: Breeding and Young Animal Rearing, Milking, and Feeding . . . 69
5.1 Reproductive Management . . . 70
5.1.1 Management of Reproduction . . . 70
5.1.2 Assistance to the Parturition . . . 73
5.1.3 Care of the Newborn Camel . . . 74
5.1.4 Adoption Techniques . . . 77
5.1.5 Management of Breeding Males . . . 78
5.1.6 Reproductive Biotechnology . . . 80
5.2 Management of Milking . . . 83
5.2.1 Presence of the Camel Calf . . . 83
5.2.2 Pace of Milking . . . 84
5.2.3 Types of Milking . . . 85
5.2.4 Milk Ejection and Milking Quality . . . 86
5.2.5 Training Camels in Mechanical Milking . . . 88
5.2.6 Hygiene of Milking . . . 90
5.3 Rearing Young . . . 94
5.3.1 Youth Sustainability . . . 94
5.3.2 Weaning . . . 97
5.3.3 Growth Control and Measurements . . . 98
5.4 Feeding Management . . . 100
5.4.1 Feed Requirements . . . 100
5.4.2 Potential Food Resources . . . 104
5.4.3 Calculation of Ration . . . 106
References . . . 109

6 Animal Health Management and Veterinary Practices in a Camel Farm . . . 111
6.1 Restraint Techniques . . . 112
6.1.1 Head Restraint . . . 112
6.1.2 Limb Restraint . . . 114
6.2 Sedation and Anesthesia . . . 119
6.3 Clinical Examination and Specimens . . . 124
6.3.1 General Clinical Examination . . . 124
6.3.2 Blood Sampling . . . 127
6.3.3 Other Types of Sampling . . . 130
6.4 Interpretation of Analyses . . . 131
6.5 Diseases and Conditions . . . 132
6.5.1 Skin Problems . . . 134
6.5.2 External Parasites . . . 137
6.5.3 Eye and Sinus Conditions . . . 141
6.5.4 Locomotion Disorders . . . 142
6.5.5 Trypanosomosis . . . 146
6.5.6 Digestive Disorders . . . 146
6.5.7 Respiratory Problems . . . 149
6.5.8 Reproductive Pathologies . . . 150
6.5.9 Major Infectious Diseases . . . 153
6.6 Necropsy . . . 154
References . . . 156

7 Economic Management of a Dairy Camel Farm . . . 159
7.1 Animal Identification . . . 160
7.1.1 Types of Identification . . . 160
7.1.2 Implementation of a Farm Database . . . 161
7.2 Demographic Model . . . 161
7.3 Economic Profitability Assessment Model . . . 163
7.3.1 Expenditure Items . . . 163
7.3.2 The Products . . . 168
7.3.3 Final Profitability . . . 170
References . . . 171

8 Camel Products and Services: From Dairy, Meat, and Nonfood Products to Riding and Transport, Including Slaughter . . . 173
8.1 Camel Milk Processing . . . 174
8.1.1 Fermented Milk: From Traditional Product to Dairy Industry . . . 174
8.1.2 Pasteurized Camel Milk . . . 175
8.1.3 Sterilized Milk . . . 176
8.1.4 Camel Milk Yogurt . . . 177
8.1.5 Camel Butter . . . 178
8.1.6 Camel Cheese . . . 178

8.1.7 Camel Milk Powder 180
8.1.8 Other Camel Milk Products 180
8.1.9 Nonfood Processing of Camel Milk 182
8.2 Camel Meat Products 182
8.2.1 Slaughter 184
8.2.2 The Nutritional Quality of Camel Meat 186
8.2.3 Processing of Meat Products 188
8.3 Other Camel Products 190
8.3.1 Wool 190
8.3.2 Camel Leather 191
8.3.3 Manure 193
8.4 The Service Functions 194
8.4.1 Saddled Mounts 194
8.4.2 Packed or Hitched Transport 200
8.4.3 The Auxiliary of Agriculture 201
8.5 Camel Transport 203
References 206

About the Authors

Bernard Faye, veterinarian and doctor of the University PARIS XII, is an INRAE research engineer and currently emeritus expert at CIRAD. As a specialist in camel breeding and production, he brings more than 40 years of experience with this species. He is author of the "Guide de l'élevage du dromadaire" (Ed. SANOFI, 1997), of "Bergers du monde" (Ed. QUAE and E&C, 2008), and contributor to the "Dromadaire pédagogique" (Ed. CIRAD-Les Savoirs partagés, 2002). Bernard further co-authored *Camel Clinical Biochemistry and Hematology* (Ed. Springer, 2018).

Gaukhar Konuspayeva (Al-Farabi University in Almaty, Kazakhstan) is a biochemist and doctor of the University of Montpellier. She serves as a consultant for FAO and has been a visiting researcher at CIRAD. Gaukhar is a specialist in camel milk and processing. The present book has greatly benefited from her contribution on the processing of dairy products.

Cécile Magnan is a veterinary doctor and equine specialty practitioner committed to the protection of circus animals. As such, Cécile has extensive experience in the care to be given to large camelids of which she keeps some specimens on her own farm transformed into a retirement home for old circus animals. Her contribution to this volume has been decisive on the aspects of health management.

General Information on Large Camelids in the World: Origin, Taxonomy, Biodiversity, and New Settlements

1

Abstract

The camelid family includes seven species (large and small) and some hybrids and crossbreeds. The large camelids occupy cold (Bactrian) and hot (dromedary) deserts of the Old World, but new and notable adoption of camels has occurred in parts of Africa and in Western countries for the last few decades. Estimated nowadays to less than 40 million heads, the world population of camels is still growing in most of the countries, especially in Africa. Despite the description of different phenotypes and zootechnical uses, the genetic variability is relatively low due to low selection pressure and high mobility of the camel herds, leading to panmictic population. However, among one-humped camel, there seems to be a difference between African and Asian ecotypes, the latter having higher milk production potential.

Keywords

Biodiversity · Camel demography · Geographical expansion · Hybridization

1.1 Origin, Taxonomy, and Present Status of Large Camelids in the World

The large camelid family originated in North America about 40 million years ago. From this nucleus of origin, two migrations – one to South America via the Isthmus of Panama and the other to the Asian continent via the Bering Strait – gave rise to the two major branches of the camelid family today, the small Andean camelids (*Lamini* group) and the large camelids (*Camelini* group), respectively. The divergence between these two groups occurred 11 million years ago in North America. The ancestor of the small camelids, the *Palaelama*, would have arrived in the Andean mountains between 1.8 million years and 11,000 years before today; the ancestor of the large camelids (*Camelops*?) would have migrated toward Asia approximately

B. Faye et al., *Large Camel Farming*, https://doi.org/10.1007/978-94-024-2237-5_1

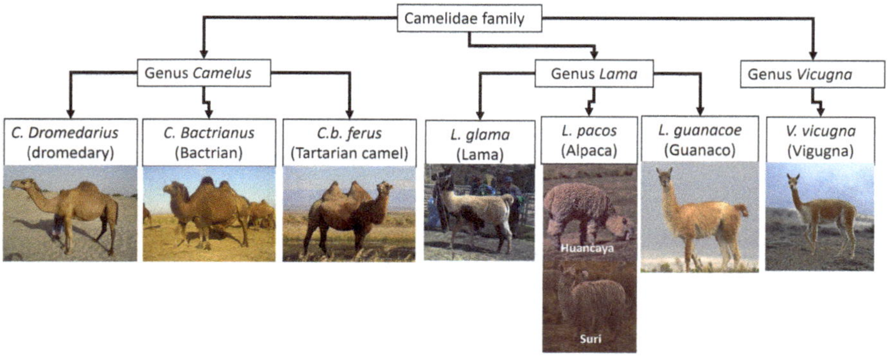

Fig. 1.1 Taxonomy of the members of the Camelidae family. (**a**) Dromedary camel (*Camelus dromedarius*). (**b**) Bactrian camel (*Camelus bactrianus*). (**c**) Wild camel of Tartary (*Camelus b. ferus*). (**d**) Lama (*Llama glama*). (**e**) Alpaca (*Llama pacos*)- Huacaya breed. (**f**) Alpaca (*Llama pacos*)- Suri breed. (**g**) Guanaco (*Llama guanacoe*). (**h**) Vicugna (*Vicugna vigugna*)

8 million years ago. The large camelids in turn split into the dromedary (one-humped camel) and Bactrian (two-humped camel), 4–5 million years ago, the former migrating to the warmer lands of the Arabian Peninsula and the latter to the colder lands of Central Asia. A final divergence occurred less than a million years ago between the Bactrian camel and what some call the Tartarian camel (Hare 1999), the remnant of which is the present wild camel.

This long history has thus structured the current family of camelids, which today comprises three genera and seven species (Fig. 1.1). The genus *Camelus* thus comprises three species: dromedary (*C. dromedarius*), also called the Arabian camel or one-humped camel; Bactrian (*C. bactrianus*), or two-humped camel, sometimes called the Asian camel; and Tartarian camel (*C.b. ferus*).

The Tartarian camel (*C. bactrianus ferus*), long considered to be a wild Bactrian camel (ancestor of the present-day Bactrian), has recently been recognized as a different species following molecular genetic studies showing a clear divergence with a full genotype. It is therefore a cousin and not a direct ancestor of the Bactrian.

Among the small Andean camelids, the genus *Lama* includes the llama (*L. glama*), properly called the alpaca (*L. pacos*) and the guanaco (*L. guanicoe*). The vicuña (*V. vicugna*), a wild species, belongs to the genus *Vicugna*, of which it is the only representative. However, some classifications include the alpaca in the genus *Vicugna* (*V. pacos*), considering that this species is the domestic version of the vicuña, and the llama, that of guanaco.

Domestication would have occurred for the Bactrian, 5000 to 6000 years ago, probably in a more western area than previously thought, i.e., toward Uzbekistan and present-day western Kazakhstan, rather than toward Mongolia. The name "Bactrian" comes from the region (a former kingdom conquered by Alexander the Great) that lies between Afghanistan, Iran, and present-day Kazakhstan. The domestication of the dromedary would be more recent (3000 to 4000 years) and, in all probability, would have taken place in the southeast of the Arabian Peninsula (present-day

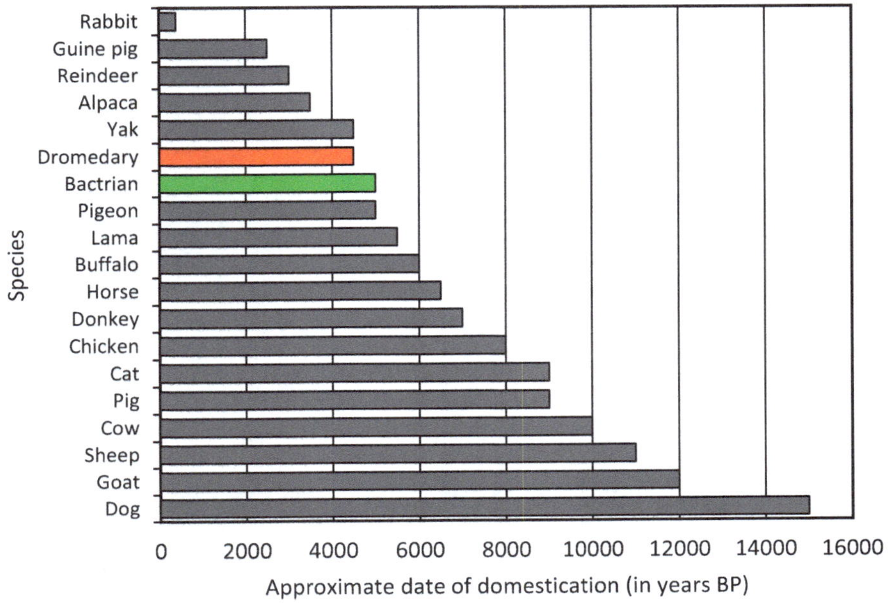

Fig. 1.2 Approximate dates of animal domestication

Sultanate of Oman, United Arab Emirates, and southern Saudi Arabia). According to current data, large camelids are therefore among the last major species to be domesticated by humans (Fig. 1.2).

The camelid family has, however, expanded since domestication with the creation of crossbreeds and hybrids. Formally, the crossing of the dromedary and the Bactrian is a crossbreeding rather than a hybridization, as both belong to the same genus (in the same way as the crossing of the zebu and the cow). Nevertheless, in everyday language, we speak of hybrids when these two species are crossed. This practice, common since ancient times along the trade routes of the Asian continent ("Silk Roads"), consisted in obtaining hybrids combining the strength of the Bactrian and the endurance of the dromedary, qualities that were very useful in the caravan activity. Today, hybridization is actively implemented in two contexts: (i) in Central Asia, particularly in Kazakhstan, to obtain females producing more milk than the parents and milk richer in fat, and (ii) in Turkey, to obtain males renowned in "camel wrestling" festivals (Fig. 1.3). Indeed, Kazakh (Faye and Konuspayeva 2012) and Turkish (Dioli 2020) breeders have implemented several hybridization schemes depending on whether the male is one- or two-humped. Note that the F1 hybrid (50% Bactrian/50% dromedary) is never crossed with also an F1 but always with a pure dromedary or pure Bactrian resulting in an F2 (25% dromedary/75% Bactrian or 75% dromedary/25% Bactrian). Indeed, the F1/F1 cross gives an animal called *Jarbaï* in Kazakhstan, impossible to manage because of a very difficult and often dangerous character. The same pattern is repeated for F3 and F4. In the end, we

a b

Fig. 1.3 Examples of hybrids of large camelids. (**a**) F2 hybrid for dairy production (Kazakhstan). (**b**) F1 hybrid for camel wrestling (Turkey)

a b

Fig. 1.4 Examples of small camelid hybrids. (**a**) *Pacovigugna* (vicuna/alpaca hybrid) and (**b**) *Wakisso* (llama/alpaca hybrid)

obtain a varied range of individuals distinguished by the shape of the hump (more or less subdivided) and the distribution of the fur on the body.

Hybridization also concerns small camelids. Hybridization between alpacas and vicuñas (*pacovigugna*) is intended to produce an individual that produces very high-quality wool, as the vicuña has an exceptionally fine fleece, although it is not very abundant (Fig. 1.4a). However, such hybridization is questioned by those who want to maintain animal biodiversity, as the vicuña is a wild species. There is also a llama/alpaca hybrid, called *wakisso* (Fig. 1.4b), which produces a larger individual with an abundant fleece. As these are two domesticated species belonging to the same genus, it is also a crossbreeding.

Finally, for completeness, mention should be made of the dromedary-llama (*cama)* hybridization trials carried out by artificial insemination at the Camelid

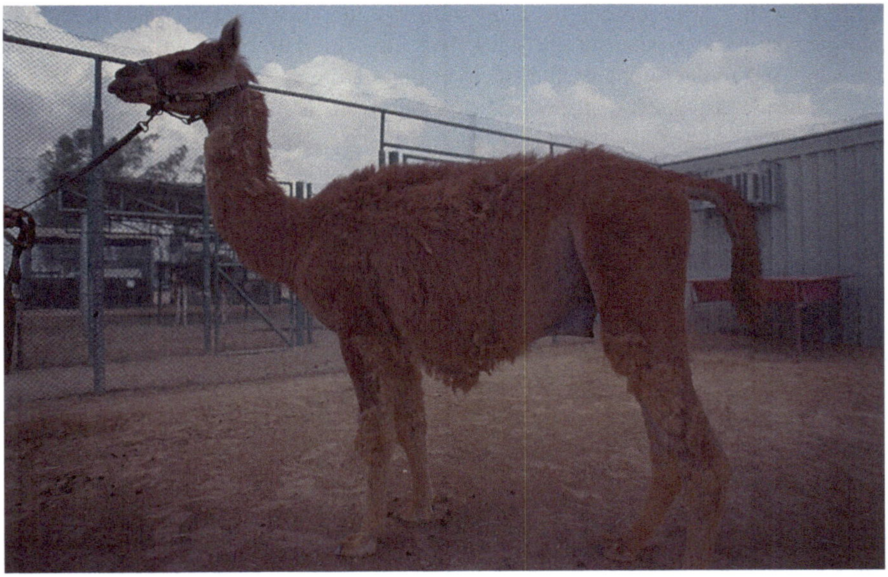

Fig. 1.5 Cama (llama/dromedary hybrid)

Breeding Centre in Dubai (Fig. 1.5). This is a true hybridization. It should be noted that dromedary/Bactrian crossbreeds are fertile, as are the *pacovigugna* or the *wakisso*, but not the *cama*.

The number of large camelids in the world is difficult to know. On the one hand, it is a population widely scattered in areas with low human density, and on the other hand, these animals are not subject to mandatory vaccination campaigns, as are cattle. In most countries without an identification and registration system, the authorities are forced to make estimates that turn out to be greatly underestimated, as shown by the few examples of census. This is the case of Chad, which estimated its camel population at 1.55 million head in 2014 and then officially declared 6.413 million head in 2015.

According to the Food and Agriculture Organization (FAO) website, the world population of large camelids is 39.5 million in 2021 (the latest year for which statistics are available). As these data do not include figures for a large number of countries with small numbers (notably animals in Western countries and some African countries outside the Sahara-Sahel region) and as they are based on estimates in more than half of the cases, it is reasonable to consider that the world population is in excess of 50 million heads, which, all things considered, is small in comparison with the 1.5 billion cattle, the 2.3 billion small ruminants, or even the 120 million horses. Finally, although official statistics do not distinguish between one-humped camels (dromedaries) and two-humped Bactrian camels, it is possible to estimate their respective shares of the world population, with the latter accounting for 100% of the Chinese, Mongolian, and Russian herds and 85% of the Kazakh herd. They are also present in small numbers in other Central Asian countries (Kyrgyzstan, Uzbekistan, Tajikistan) but also in Turkey, Iran, Afghanistan, Pakistan, and India. Globally, the Bactrian population is estimated at less than one million head, i.e., 2.8% of the world population of large camelids in 2019 (Faye 2020).

Four types of population growth by country have been described: (i) Countries that experienced a population decline until the 2000s, followed by a slight increase in numbers. This group of countries corresponds to about 5% of the world population and includes countries in Asia (China, India, Mongolia, Afghanistan), the Middle East (Iran, Iraq, Israel, Jordan, Lebanon, Kuwait, Turkey), and a few in Africa (Egypt, Libya, Senegal). In this group, India only has experienced a continued decline since the 2000s. (ii) Countries that have experienced regular growth of around 1.8%/year for the past 60 years (50% of the world population), mainly in Africa (North Africa, Mauritania, Burkina Faso, Horn of Africa –Sudan, Ethiopia, Somalia) but also in Central Asia, Pakistan, and Bahrain. (iii) Two countries that have experienced strong growth, after a brief period of decline in the 1960s and 1970s. These countries are Syria and the Emirates, which are growing at around 6% per year but account for only 1.5% of the world's population. (iv) Countries that have experienced almost explosive growth since the 2000s, after a long period of steady growth. This group includes the Sahelian countries (Mali, Niger, Chad, Nigeria, Kenya, Djibouti) and the Arabian Peninsula (Qatar, Saudi Arabia, Oman), which clearly shows that there is no link between the demographic dynamism of the camel population and economic growth.

1.2 Biodiversity and New Settlements

Camel genetics has made enormous progress in the last 20 years, particularly thanks to the contribution of molecular tools. This has made it possible to better describe the diversity of the world camel population. Previously, about 50 "breeds" had been identified worldwide, although the notion of a breed in large camelids may be questionable, as the selection pressure on a particular trait (milk or growth performance, wool quality, running ability, etc.) has been low overall. Throughout history, the great mobility of the herds and the caravan activity over long distances have favored the mixing of populations, leading to what is called a "panmictic" population, i.e., with low genetic variability.

As far as the nomenclature of breeds (or rather "ecotypes") is concerned, there is a great deal of confusion, as the same homogeneous group (especially in terms of coat color) may have different names associated with a region or tribe. Nevertheless, morphological types can be clearly distinguished, as man has historically oriented the selection according to his needs for production, transport, or racing (Fig. 1.6). Overall, one could distinguish (at least in the dromedary) the following:

- Long, tall, slender-limbed types used more as saddle animals or even racing animals

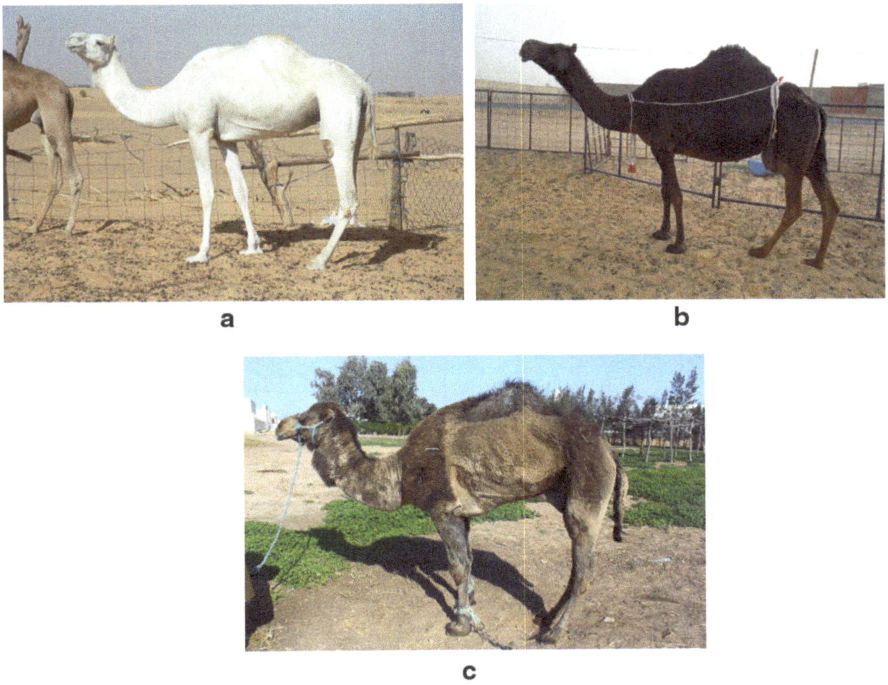

Fig. 1.6 Main morphometric types. (**a**) long linear; (**b**) mediolinear; (**c**) brevilinear

- More or less large midline types, dedicated to dairy or meat production or both
- Short, stocky, well-built types, preferably used for pack or cart transport

However, such a summary classification does not cover the diversity of types observed throughout the world. The use, certain phenotypic characteristics, color of the coat, and ecosystems of origin shape a much richer panel. In this respect, we can cite the "breeds" known in the Arabian Peninsula, the probable place of domestication of the dromedary. This is therefore where the greatest variability of the types at the origin of all the dromedaries in the world is to be found. Globally, geneticists have identified three groups (genotypes): one that originated from animals that migrated to the Middle East as well as North and West Africa; another that originated from types that populated the Horn of Africa and South Asia; and a third, occupying mountainous and coastal areas that did not migrate outside the Peninsula (Almathen 2014). To establish a classification of phenotypes, various physical measurements were proposed, such as height at withers, abdominal girth (behind the hump), body length, neck length, etc. (see Sect. 5.3.3). This approach resulted in the identification of 12 different phenotypes across the country (Fig. 1.7).

In the Bactrian camel, Chinese, Mongolian, and Kazakh researchers have also established a list of "breeds" that are distinguished mainly by the quality of their fleece (Fig. 1.8).

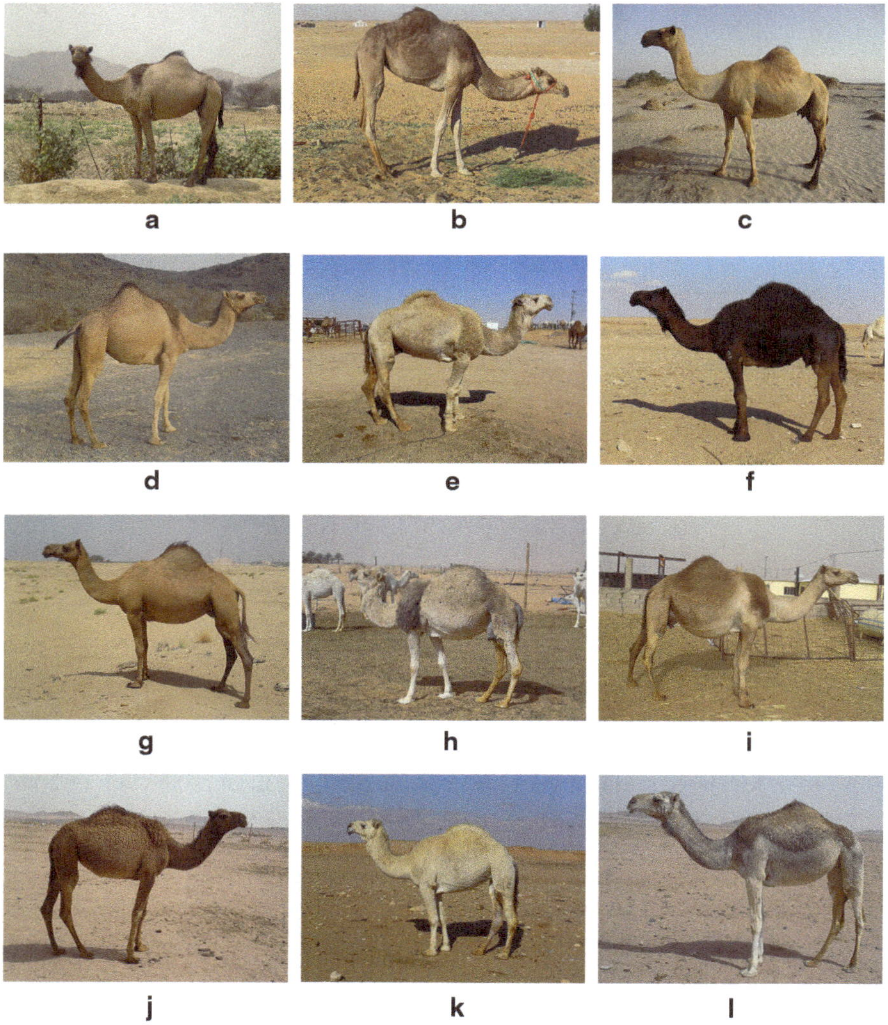

Fig. 1.7 Diversity of camel phenotypes in the Arabian Peninsula. (**a**) Awadi. (**b**) Asail. (**c**) Awarik. (**d**) Hadhana. (**e**) Homor. (**f**) Majaheem. (**g**) Saheli. (**h**) Shageh. (**i**) Shaele. (**j**) Sofor. (**k**) Waddah. (**l**) Zargeh. (Source: Faye et al. 2011)

However, in the absence of clearly established international rules on the parameters to be considered for phenotypic analysis, and with the low number of studies associating genotypes and performance control, knowledge of genetic variability in this species remains modest. Several international projects are underway to address these issues.

The low selection pressure has also resulted in the absence of a breed which, because of its exceptional qualities (like the Holstein cow in the dairy sector), was in danger of becoming invasive, even if phenotypic studies, supported by genotypic

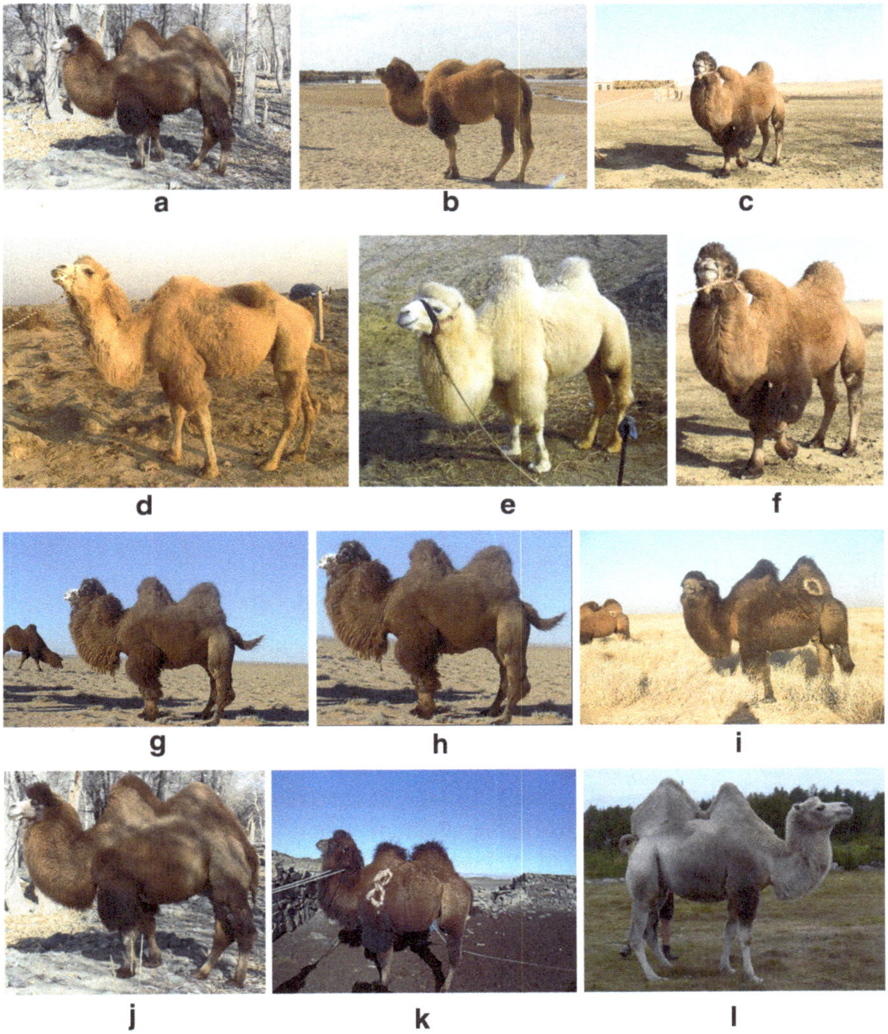

Fig. 1.8 Diversity of Bactrian phenotypes in China (Ch), Mongolia (Mo), Russia (Ru), and Kazakhstan (Kz). (**a**) Jungar (Ch). (**b**) Qinghaid (Ch). (**c**) Sonid (Ch). (**d**) Tarim (Ch). (**e**) Alxa (Ch). (**f**) Galbiin Gobiin Ulaan (Mo). (**g**) Heniin Hetsiin Huren (Mo). (**h**) Jungar (Ch). (**i**) Tokhom-Tungalag (Mo). (**j**) Astrakhan (Ru). (**k**) Crimean (Ru). (**l**) Kalmuk (Ru). (**m**) Kzylordinskii (Kz). (**n**) Uralobukeevskii (Kz). (**o**) YuzhnoKazakhstanskii (Kz). (Photos B. Faye, G. Konuspayeva and S. Hasi)

studies, clearly distinguish Asian populations – which are larger and have better milk production – from African populations. This absence allows a certain diversity to be maintained, which is important to preserve. Currently, only one camelid species is the subject of a conservation program, the wild Tartarian camel, because of its small numbers (less than 2000 heads) and the risk of hybridization with the Bactrian camel.

Fig. 1.8 (continued)

Although in the FAO database only 48 national entities officially declare a population of large camelids (Fig. 1.9), dromedaries and Bactrians are effectively present in many other countries. These new settlements concern (i) African countries on the fringes of the Sahara, which are benefiting from the current expansion of camel farming following the recurrent droughts that have affected the continent (e.g., Cameroon, Uganda, Tanzania, etc.); (ii) the desert countries of southern Africa (Namibia, Botswana); and (iii) the western countries (USA, Europe, and Australia). In this last group of countries, large camelids are mainly raised for tourism, but their integration into agricultural systems (milk, meat) is increasing, as shown by the emergence of dairy farms in Holland, Germany, France, Spain, and the USA (Faye 2022).

Genetically, the populations in these new settlements are often of varied geographical origin and are the result of uncontrolled interbreeding between different types or even species. One study has shown the importance of Bactrian genes in European camel populations (Fig. 1.10).

Finally, we should note the particularity of the Australian camel population. The camels were imported during the nineteenth century (mainly from Afghanistan and what is now Pakistan) and were used for agricultural and transport activities in the desert heartland of Australia until the 1920s and 1930s. More or less abandoned in the bush after the motorization of agriculture, the dromedaries became feral (marooned to some degree) and multiplied to the point of posing a real environmental problem. Estimates of the camel population range from 400,000 to 1 million (Faye 2020).

Fig. 1.9 Countries officially reporting a large camelid population on the FAOstat website

Fig. 1.10 Dromedary farming in northern France

References

Almathen F (2014) Genetic diversity and demographic history of dromedary camel (*Camelus dromedarius*). PhD thesis, University of Nottingham, UK

Dioli M (2020) Dromedary (*Camelus dromedarius*) and Bactrian camel (*Camelus bactrianus*) crossbreeding husbandry practices in Turkey and Kazakhstan: an in-depth review. Pastoral Res Policy Pract 10(6) [en ligne]. https://doi.org/10.1186/s13570-020-0159-3

Faye B (2020) How many large camelids in the world? A synthetic analysis of the world camel demographic changes. Pastoral Res Pol Pract 10(25) [en ligne]. https://doi.org/10.1186/s13570-020-00176-z

Faye B (2022) Is the camel conquering the world? Anim Front 12(4):8–16. https://doi.org/10.1093/af/vfac034

Faye B, Konuspayeva G (2012) The encounter between Bactrian and Dromedary Camels in Central Asia. In: Knoll E-M, Burger P (eds) Camels in Asia and North-Africa- interdisciplinary perspectives on their past and present significance. Austrian Academy of Sciences Press, Vienna, pp 27–33

Faye B, Abdallah H, Almathen F, Harzallah B, Al-Mutairi S (2011) Camel biodiversity. Camel phenotypes in the Kingdom of Saudi Arabia. Camel Breeding, Protection and Improvement Center, project UTF/SAU/021/SAU, FAO publ., Riyadh (Saudi Arabia), 78 p

Hare J (1999) The lost camels of Tartary. A quest into forbidden China. Little, Brown Book Group Publ., New York, p 256

Anatomical Features of Large Camelids

2

Abstract

Large camelids have a remarkable external anatomy characterized by the presence of a long neck, a horizontal head, elongated limbs, and particularly the one or two humps on the back. If the camel's skeleton does not have a special structure, the dentition is unique in the presence of caniniform teeth; wearing of adult teeth can indicate approximate age of the animal. Regarding internal anatomy, we will remember the presence of isolated lymph nodes, the nonlobulated shape of the kidney and lungs, the absence of a gallbladder, the helical part of the intestine, and the flexible shape of the foot. Stomach and genital organs will be described in further chapters.

Keywords

Morphology · Dental formula · Internal organs · Abdominal topography

The general morphology of the large camelids is sufficiently particular to have intrigued zoologists since Aristotle. With its hump (or humps in the case of the Bactrian) on its back, which is very characteristic of the species, its disproportionately long neck and its head held horizontally, which gives it the appearance of being mounted on a pedestal, the dromedary deserves to be studied for its anatomical peculiarities, which make it the largest animal adapted to desert life. Although there are some variations in this general morphology by virtue of geographical differences, it is notable that anatomical plasticity is less spectacular than in other domestic species. The fact that the dromedary is ecologically specialized to a biotope marked by aridity probably contributes to limit its phenotypic variability, whereas diffusion in very different ecosystems seems to authorize greater morphological differences in species such as cattle or small ruminants that have invaded almost all the ecosystems of the planet.

Overall, what are the anatomical features to remember?

B. Faye et al., *Large Camel Farming*, https://doi.org/10.1007/978-94-024-2237-5_2

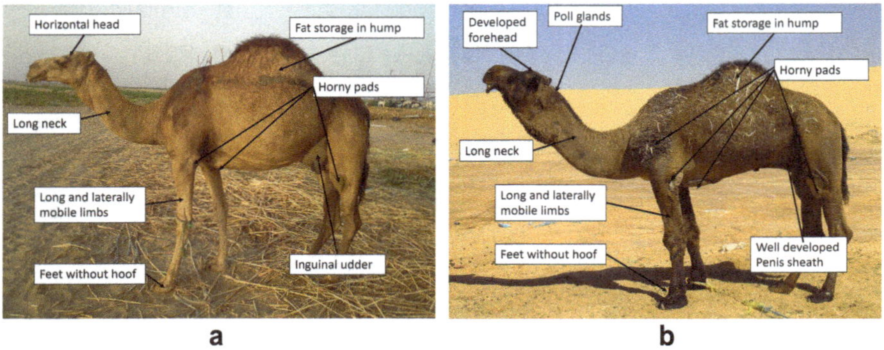

Fig. 2.1 Morphological characteristics of female (**b**) and male dromedaries (**a**)

2.1 General Compliance

The general appearance of the large camelids is easily recognizable. The concentration of fat reserves on the back, forming one (for the dromedary) or two humps (for the Bactrian), is probably their most emblematic anatomical feature. But other points are worth highlighting in the dromedary (Fig. 2.1):

- The head, horizontal, parallel to the ground, unlike other herbivores, located at the end of a long neck forming an open arc. Lips very mobile, the upper lip split, and the lower sometimes hanging. Ears small, more or less rounded
- The long, slender limbs, very mobile laterally, the attachments to the body being loose
- Feet without hooves, but with nails at the end of each phalanx (two on each foot)
- Horny pads on the sternum, knees, and elbows, located at the areas of friction with the ground when the animal is in a "stooped" position (sitting)
- In the female, an inguinal udder similar in appearance to that of the cow (four quarters, four teats)
- In males, the well-developed penile sheath
- The male's head often having a well-developed forehead and, at the back of the skull, occipital glands that are very active during the rut

In the Bactrian (Fig. 2.2), other peculiarities can be highlighted:

- A shorter head with a more pronounced forehead, including in females
- Abundant fur on the throat, under the neck, elbow, and sometimes the shoulders (and abundant winter fleece)
- An abundant mane in males
- Two humps that hang down like empty sacks when the fat is depleted (unlike the dromedary's hump which "melts") then stand up when it is replenished

Fig. 2.2 Morphological characteristics of female (**a**) and male (**b**) Bactrians

- A long and stocky neck, with a more marked inflection than in the dromedary
- An inguinal udder (F) and a penile sheath (M) less developed than in the dromedary

Hybrids (Fig. 1.3) combine the morphological characteristics of both parents, the main feature being the more or less marked fusion of the two humps.

2.2 The Skeleton

The dromedary skeleton is composed of thick, heavy bones representing about 20% of the carcass weight. Its general morphology is similar to that of other large mammals (Fig. 3.3). The skull is similar in size to that of the horse and has a very prominent occipital crest, to which is attached a cervical ligament strong enough to support a heavy head at the end of a long neck. The sinuses are wide and deep. The bony part of the soft palate is narrow, leaving room for a soft part that can be easily exteriorized in males in rut (see Sect. 3.2). The long lower jaw has a marked central constriction, which makes it fragile and leads to frequent fractures during occasional fights between males.

Despite its length, the neck of large camelids has only seven cervical vertebrae (like most mammals). The rest of the neck differs little from other domestic herbivores: 12 thoracic vertebrae (13 in cattle and sheep), 7 lumbar vertebrae (6 in cattle and sheep), and 4 sacral vertebrae (5 in cattle and 4 in sheep). As the hump is only a fatty mass, unlike the musculo-fatty hump of the zebu, the spinous processes of the thoracic and lumbar vertebrae do not have additional muscle attachments, and their length is therefore not affected. The bones of the limbs are long, allowing the abdominal and thoracic organs to be kept away from the ground when standing, thus avoiding the effects of a too cold or too hot ground (Fig. 2.3).

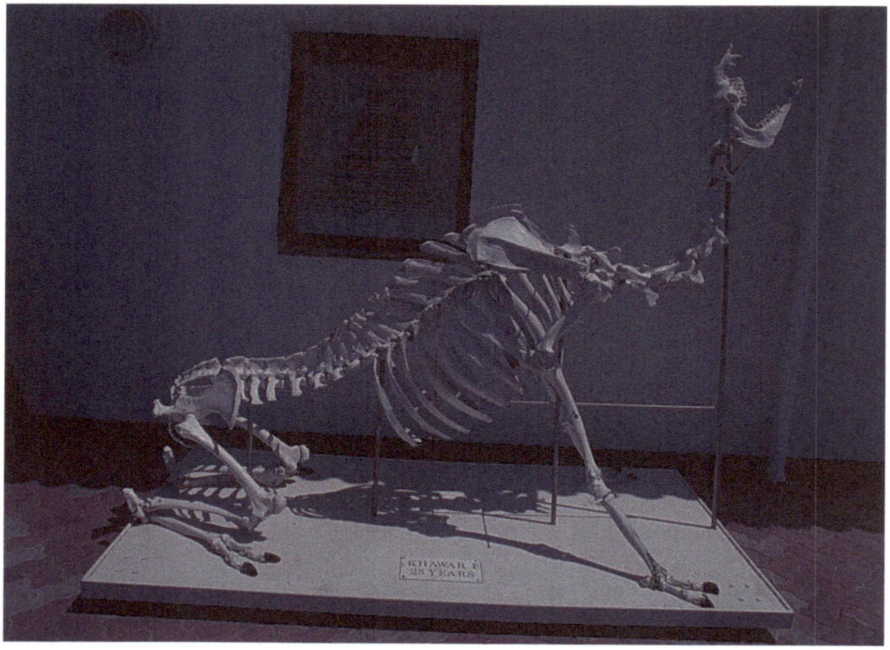

Fig. 2.3 Skeleton of an adult camel, Central Veterinary Research Laboratory (CVRL), Dubai

Table 2.1 Dental formulae of large young ("deciduous") and adult camelids

Jaw	Incisors			Canines	Premolars			Molars		
Young	I1	I2	I3	C	PM1	PM2	PM3	M1	M2	M3
Sup.	0	0	1	1	1	1	1	0	0	0
Inf.	1	1	1	1	1	1	0	0	0	0
Adult	I1	I2	I3	C	PM1	PM2	PM3	M1	M2	M3
Sup.	0	0	1	1	1	1	1	1	1	1
Inf.	1	1	1	1	1	1	0	1	1	1

2.3 Dentition

As with most mammals, camelids have a temporary dentition (the "milk teeth"), which disappears in the adult, and a permanent dentition which gradually replaces it from the age of 1 year (eruption of the first molars). The juvenile dental formula consists of 22 teeth, and the adult dental formula consists of 34 teeth in total (Table 2.1).

Large adult camelids differ from domestic ruminants in having a pair of "caniniform" incisors in the upper jaw, a pair of canines in each jaw, three premolars in the upper jaw, and only two in the lower jaw (Fig. 2.4). The first premolar is

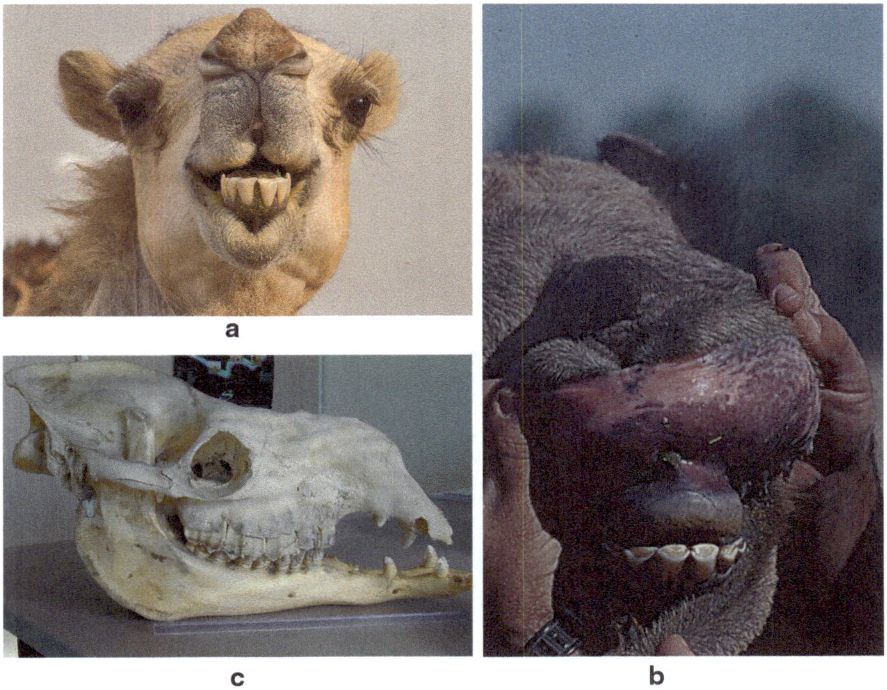

Fig. 2.4 Permanent dentition of an adult camel

isolated from the others and, like the upper incisor, simulates an additional canine tooth ("caniniform" premolar).

The evolution of the dental formula and incisor wear in adults is an empirical method for estimating the age of the animal. However, individual differences and differences in health and environmental conditions (especially sandy environments that accelerate tooth wear) lead to variability and obvious caution in interpretation (Fig. 2.5).

The gap between the lower incisors is a good criterion for distinguishing older animals. A gap between the incisors is considered to occur in animals over 15 years of age. The wear of the teeth is a factor of longevity. Indeed, with rapid wear (sandy environment), it is not common to find animals older than 20 years. In good conditions, large camelids can live beyond 30 years.

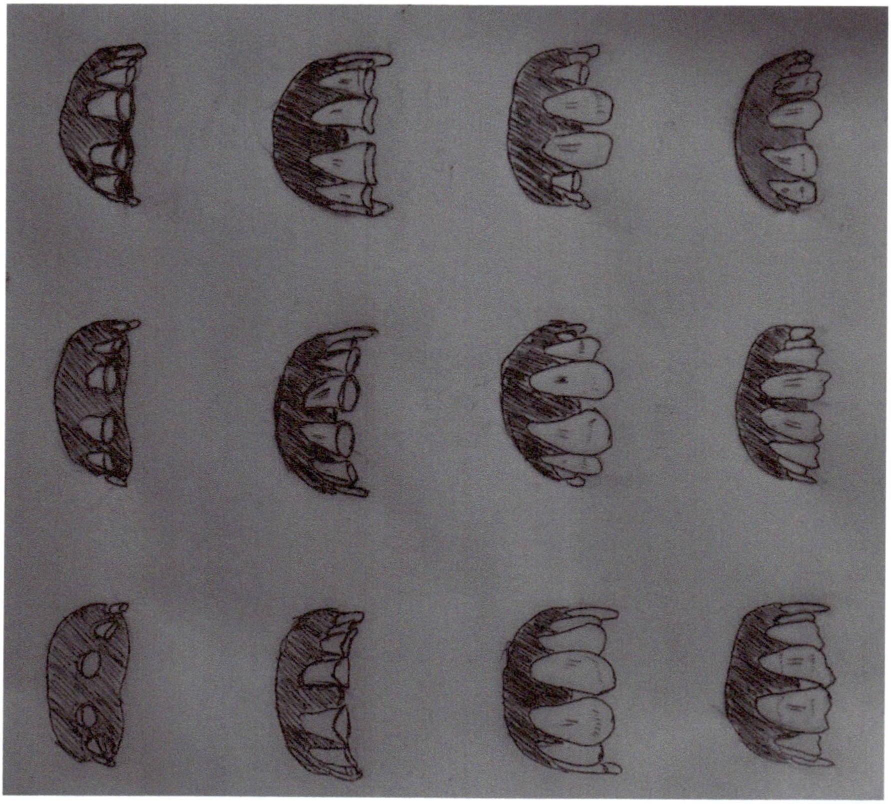

Fig. 2.5 Assessment of age based on dental status. At 3 years of age, the temporary mouth is done (BFt), and at 7 years of age, the permanent mouth is done (BFp)

2.4 Internal Anatomy

The genital organs and stomachs will be described in the chapters devoted to the physiological basis of reproduction (Chap. 3) and digestion (Chap. 4). The focus here will be on the internal features of camelids and the visceral topography useful for surgical procedures.

2.4.1 Muscular Apparatus

Muscularly speaking, three things should be noted: (i) despite its length, the neck is not particularly muscular, which makes it possible to control the animal during restraint by folding the neck over the thorax; (ii) there is an extensor muscle common to both toes in addition to a muscle for each of them; (iii) there is no skin muscle:

large camelids cannot therefore "vibrate" their skin to get rid of flies as in the horse or the cow. For the rest, the structure of the muscles is similar to that of other domestic species. The muscular system accounts for about 70% of the camel's body mass.

2.4.2 Nervous System

The dromedary's brain is comparable to that of the horse, both in terms of morphology and volume. It has a marked cleft between the two cerebral hemispheres, a well-developed temporal lobe, and olfactory bulb. The cortex is covered with irregular convolutions. Available data report that the total weight of the brain in the dromedary is 680 g, its length is 15 cm, and its volume is 446 cm^3. The cortical surface area is 3222 cm^2, and the thickness of the cortex is 2.3 mm (Fig. 2.6).

The surface of the cerebellum is covered with finely spaced parallel grooves, in striking contrast to the large irregular convolutions of the cerebral cortex.

2.4.3 Lymphoid System

The characteristics of the lymphatic system of large camelids are the relatively small number of lymph nodes and some unusual locations such as the external thoracic lymph node or the lower cervical lymph node. Also, unlike other species, the lymph

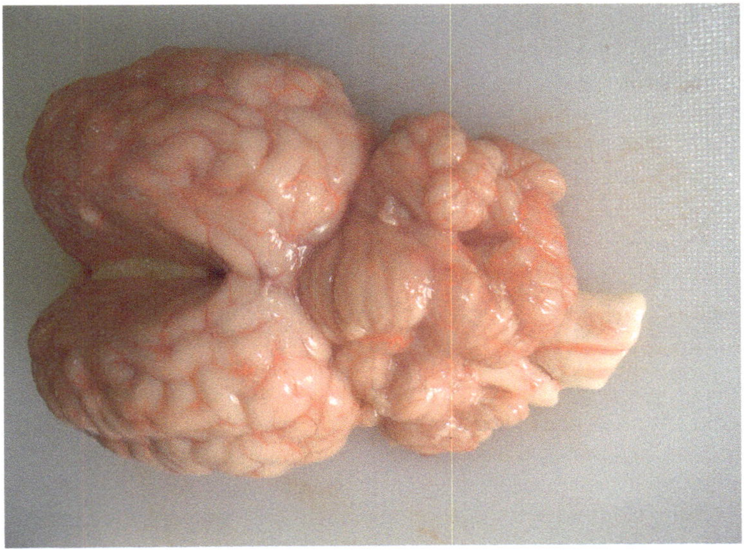

Fig. 2.6 Dromedary brain showing the two hemispheres with its marked fissure, the cerebellum, and the medulla oblongata. (Photo R. Seboussi)

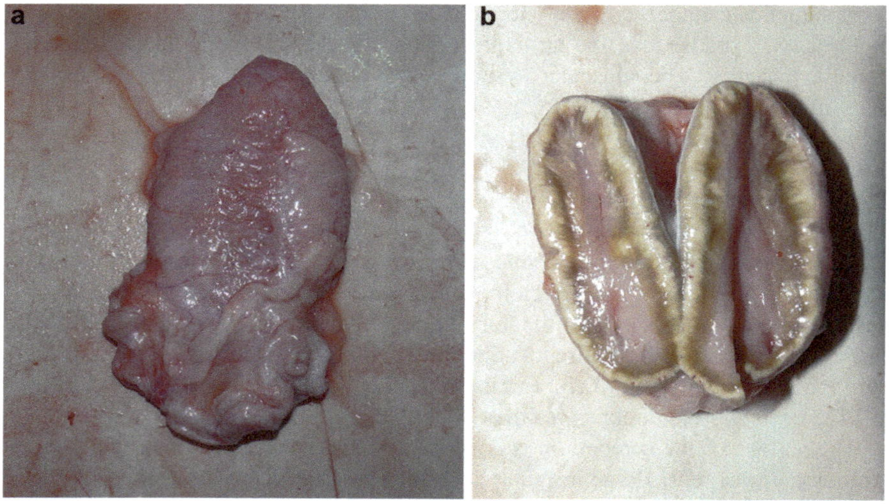

Fig. 2.7 Lymph node: (**a**) general view; (**b**) longitudinal section view

nodes of the camel do not occur in groups but as isolated organs. The anatomical position of the main palpable surface lymph nodes can be found in Chap. 6 ("Health Management"). The particularity of these nodes is their lobulated shape (Fig. 2.7), which may lead to confusion with salivary glands, and their hardness due to a high content of fibrous tissue.

These lymph nodes, which act as filters or traps for foreign particles, have important clinical significance in the camel, as they abscess readily when the animal is affected by "abscess disease" (see Chap. 6, "Health Management").

Two important organs in the animal's immune defense system belong to the lymphoid system: the spleen and the thymus. In the camel, the spleen is semi-lunar in shape, and in cross-section, it is comma-shaped (Fig. 2.8).

The spleen is located above the dorsal sac of the first compartment (rumen) and weighs 1.5 to 1.6 kg. The thymus, placed in front of the heart, is only visible in the young, this organ disappearing a few months after birth.

2.4.4 Blood System

The blood volume in the camel is higher than in other species: it averages 93 mL of blood per kilogram of body weight compared to about 60 mL in sheep and cows and 75 mL in goats and horses. The total volume of blood is therefore about 35–55 L of blood in an adult camel, depending on its weight. This volume is related to the animal's ability to manage its internal temperature according to the external environment, as blood plays an essential role in thermoregulation. However, during bleeding, only part of this blood can be collected (about 1/25 of the animal's weight, i.e., 15–25 L).

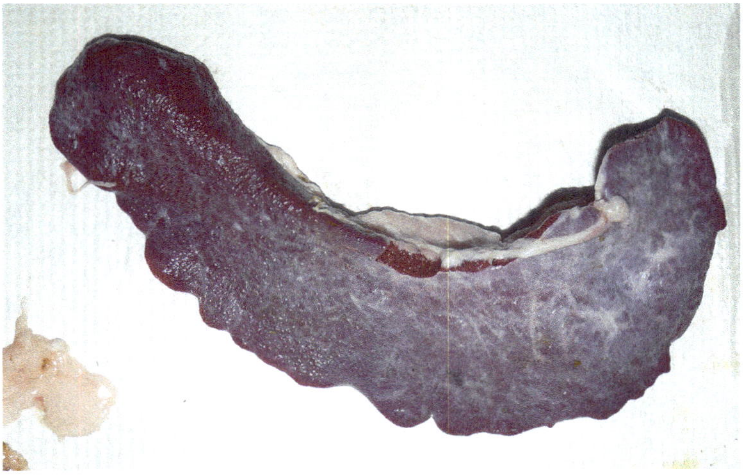

Fig. 2.8 General view of the upper part of a dromedary spleen

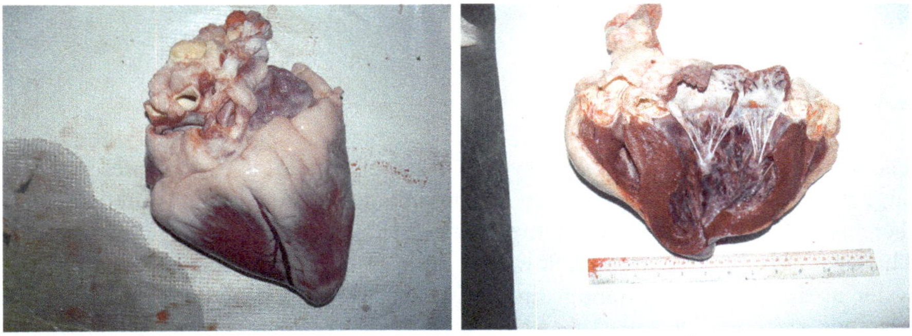

a b

Fig. 2.9 Camel heart: (**a**) external view; (**b**) longitudinal sectional view

The heart of the camel is structurally and functionally indistinguishable from other species. Its weight is about 1.5 kg (Fig. 2.9). The heart rate of the camel averages 45.6 beats/minute with a normal range of 30–57 beats/minute. In the racing camel, the heart rate accelerates to 140–150 beats/min with a rapid recovery time of about 30 minutes but with a decrease of more than 40% in the first minute after running.

The jugular vein is large and easily visible near the head (Fig. 2.10), making it relatively easy to draw blood from the distal part of the neck (see Sect. 6.3). In contrast, slaughter is performed by severing the jugular vein and carotid arteries in the proximal part of the neck, near the entrance to the chest.

The pulse can be felt on the posterior tibial artery, even with the animal in the prone position, although this is a rarely used measurement.

Fig. 2.10 Jugular vein and nostrils

2.4.5 Respiratory System

The particularity of the respiratory system of large camelids is first the presence of a large nasal cavity and sinuses subdivided into numerous convolutions. In particular, there is a lateral blind sinus sac in the sinus, which is not found in any other species. This anatomical configuration allows the dromedary to recover a large part of the water at the time of expiration through the nasal passages. Moreover, the nasal passages are connected to the outside by nostrils that can be completely closed (Fig. 3.10), thus avoiding drying of the nasal mucosa and maintaining a humid atmosphere in the upper respiratory tract that limits water loss.

The lungs of camelids are lobeless, unlike those of cattle, but there are no notable structural differences (Fig. 2.11). Each main bronchus divides into an apical bronchus, a cardiac bronchus, and the large diaphragmatic bronchi. No cartilage or glands are present in the dromedary bronchioles. The respiratory bronchioles are absent. The respiratory rate of the dromedary is 13 to 16 breaths/minute. This rate can vary depending on the season (and therefore the climate).

2.4.6 Digestive System

The esophagus of the camel is a long muscular tube of great capacity, which carries the chewed food bolus from the larynx to the stomach. Because of the length of the neck, this organ is also long, measuring between 1.65 and 2.15 m depending on the individual. The internal mucous membrane is lined with glands that secrete an

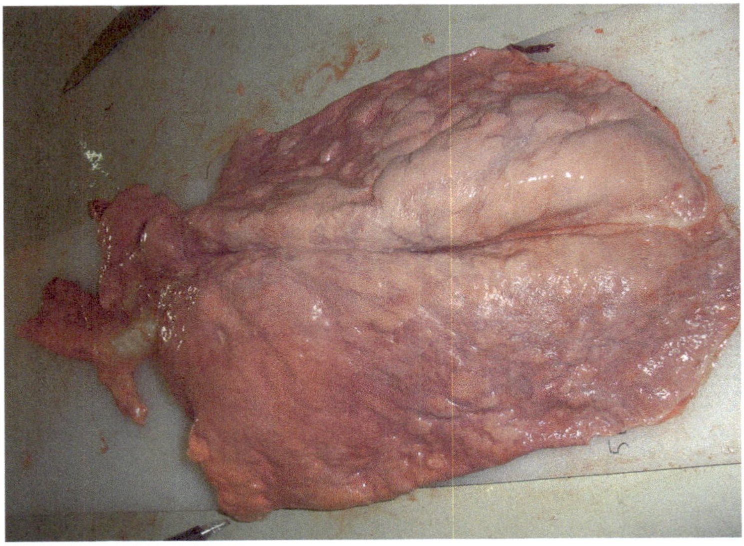

Fig. 2.11 Dromedary lungs. Note the absence of lobes. (Photo R. Seboussi)

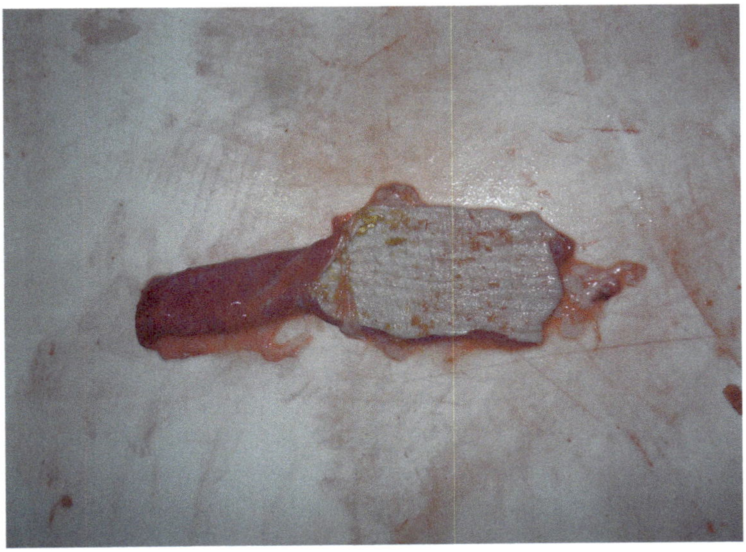

Fig. 2.12 Cross-section of a portion of the dromedary esophagus showing the striated mucosal layer on the inner surface

abundant mucus to facilitate the passage of fodder by lubricating it. In the larynx, the dromedary's esophagus has a thick layer of mucous membrane that is striated lengthwise (Fig. 2.12).

Fig. 2.13 Overview of the anatomical parts of the gut of large camelids

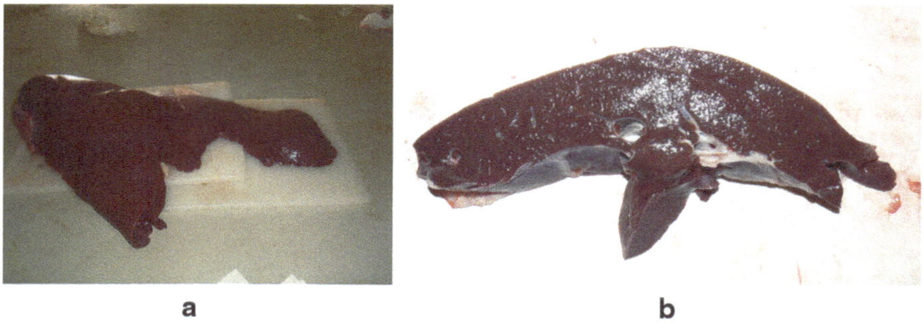

Fig. 2.14 Dromedary liver: (**a**) parietal view; (**b**) cross-sectional view showing the parietal side above and the visceral side below

The stomachs are described in the section on the physiological basis of digestion (Chap. 4). The intestinal part of the digestive tract is unremarkable. The length of the small intestine in adults is about 40 m, and that of the large intestine is about 20 m. The main feature to note is the particular configuration of the colon, as it has a helical section, which allows for increased water reabsorption from the intestinal contents and, as a result, the emission of fairly hard, very poorly hydrated fecal pellets (Fig. 2.13).

The liver, which weighs about 7 kg in adults, lacks a gallbladder, and the common bile duct is common to the single pancreatic duct. The camel liver is triangular with 4 main lobes and two sides, parietal (Fig. 2.14) below the diaphragm, and visceral above the intestines to which it is attached by various ligaments.

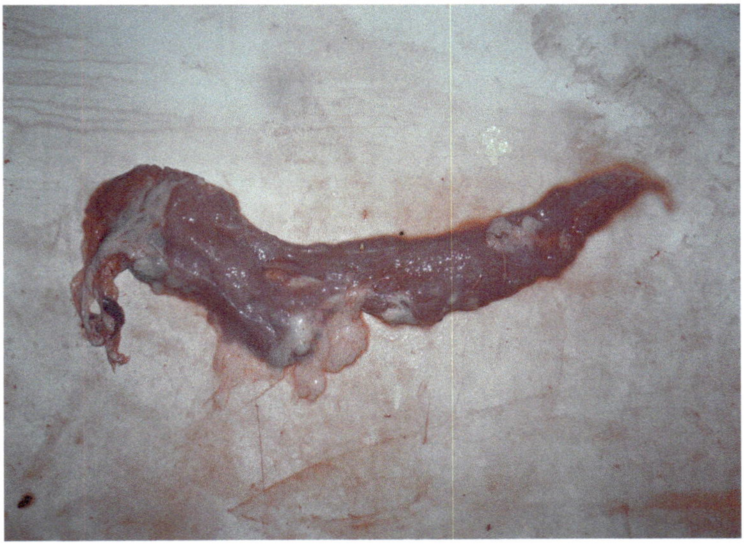

Fig. 2.15 Dromedary pancreas

The pancreas has an indefinite shape and is located between the stomach and the duodenum. It weighs about 170 g. It is largely covered with fat (Fig. 2.15).

2.4.7 Urinary System

The kidney in the dromedary is probably the most central organ in water regulation and the reabsorption of many metabolites and electrolytes, especially during the rapid dehydration/rehydration cycle. It also plays an important role in the recycling of dietary nitrogen (see Chap. 4 on the urea cycle). The left kidney is located below the lumbar vertebrae, just above the rumen, near the transverse processes of the 5^{e}, 6^{e}, and 7^{e} lumbar vertebrae, while the right kidney is placed before the left, in the hollow of the flank between the 12^{e} rib and the transverse process of the first two lumbar vertebrae.

The volume of the dromedary kidney is about 850 cm^3, and its weight is about 1 kg. The shape of the kidney is very different from that of cattle: it is brown, bean-shaped, and, above all, has no lobes like the human kidney (Fig. 2.16). Its internal structure is also distinguished by the remarkable thickness of the medulla (the part below the outer cortex), which shows the presence of a large number of nephrons (the functional units of the kidney, in the form of tubules) and especially the very long "coves of Henle," the curved part of these tubules, which contributes to the ability of camelids to excrete highly concentrated urine during dehydration. It has been calculated that the tubules, as a whole, represent an average exchange surface

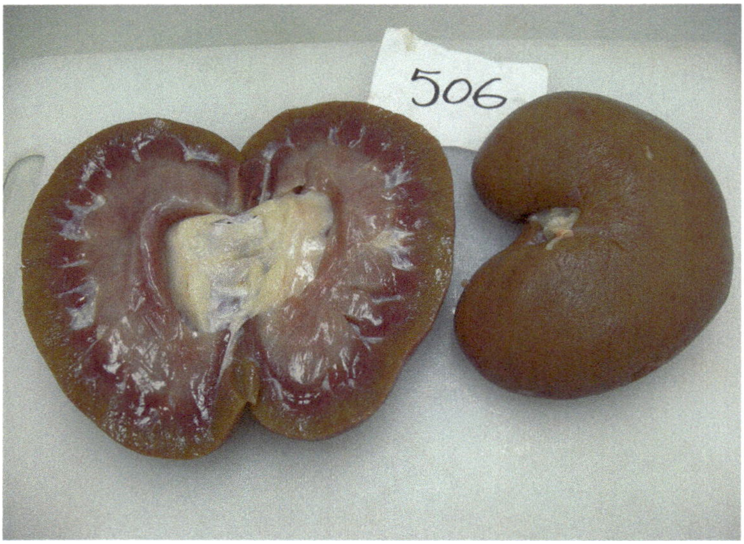

Fig. 2.16 Left kidney of dromedary in general view and longitudinal section. (Photo R. Seboussi)

of 9.5 m^2, which explains the animal's ability to reabsorb excreted water and recover urea.

The ureters, which transfer urine from the kidneys to the bladder, measure 50–55 cm for the right kidney, which is further forward, and 35–40 cm for the left kidney.

The bladder is small compared to the size of the animal. When empty, it is no more than 5–7 cm long and 4–5 cm wide and has a capacity of 600–700 cm^3. It is egg-shaped (Fig. 2.17), the position totally pelvic, and, unlike other domestic species, it is located dorsally to the rectum and prostate in males, or to the vagina in females.

2.4.8 Locomotor System

Although not part of the internal anatomy, the structure of the camel's foot is sufficiently particular to devote a separate paragraph to it. Indeed, the camel has no hoof, which places it in the group of digitigrades and not unguligrades. The foot is wide and elastic, with two phalanges (Fig. 2.18). Between the phalangeal bone and the sole, there is a connective ball, the size of a hen's egg for each finger, which is extremely elastic and acts as a shock absorber. This explains why, on the one hand, the foot easily follows the irregularities of the ground and, on the other hand, why the foot spreads out when the animal distributes its full weight. Unlike unguligrades, the foot is not constrained by the hoof wall, which explains the low clinical expression in

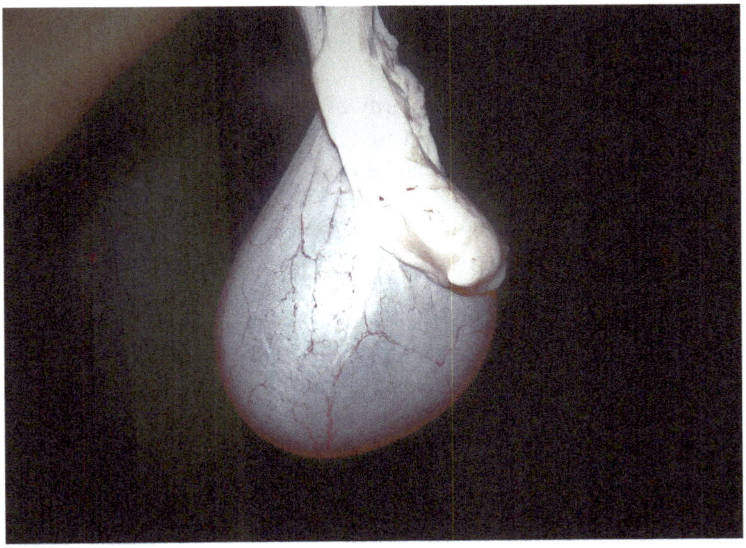

Fig. 2.17 Dromedary bladder

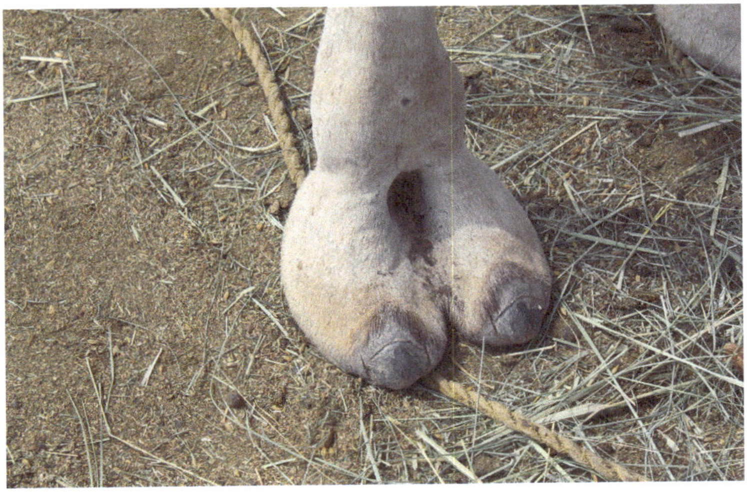

Fig. 2.18 Dromedary foot. Note the presence of a hoof at the end of the two phalanges

laminitis. Moreover, the pressure that the foot exerts on the ground is much lower than that of a cow of the same weight, 2.6 and 8.9 Newtons/cm^2, respectively.

Such a foot is therefore well suited to walking on sandy ground. It can easily be compared to a tire whose inner tube has been replaced by a fatty tissue that gives the whole a remarkable flexibility. The sole is covered by a thick, horny skin with a dermis rich in sweat glands, which maintains a certain humidity. However, its composition, if it facilitates the displacements in the dunes, makes the progression in the zones with aggressive grounds (sharp pebbles) more difficult and sometimes traumatizing.

Concerning locomotion, the particularity of the large camelids, unlike most other quadrupeds, is the amble walk with a rhythm of 38 to 43 steps per minute. The dromedary is renowned for its very regular step. It was based on this reputation that Eratosthenes, a Greek mathematician of the IIIe s. BC, was able to estimate the circumference of the Earth to within a few kilometers[1]!

This amble mode of locomotion gives the animal a characteristic undulating gait, with all the weight of the body being carried alternately on the left lateral part and then on the right lateral part. Such a gait would be characteristic of species or breeds (in very limited numbers) living in difficult areas with limited resources, as it would allow a notable saving of energy (Senault 2013). In the common animal, the trot is the normal gait. Galloping is exceptional and is hardly observed except in racing animals. At a walk, the dromedary's speed is about 6 km/h. It doubles in the trotting animal and triples in the galloping camel.

2.4.9 Abdominal Topography

The visceral topography is interesting to know in large camelids in case of surgical intervention concerning vital organs (caesarean section, trocardization) or in case of medical investigation (evaluation of cardiac and respiratory rhythms, measurement of ruminal activity). For these species, the examinations are often carried out on the animal in the sternal position (decubitus sternal), so the anatomical plates are usually of the animal in the sitting position (Oushine 1989). They show in summary the following points (Fig. 2.19):

- On the left lateral part, one can have access to the left kidney pushed backward against the 5^{e}, 6^{e}, and 7^{e} lumbar vertebrae; to the spleen, situated in the hollow of the left flank, adhering to the rumen and close to the kidney; to the rumen, which occupies most of the abdominal space; and to the spiral colon, which is interposed between the caudal extremity of the rumen and the entrance to the pelvic cavity. The abomasum projects only into a very small space between the 6^{e} and 7^{e} intercostal space.
- On the right lateral part, the resection of the abdominal wall allows access to the mass of the jejunum's convolutions and in the caudal part, to the terminal colon, which ends in the rectum in the pelvic part; the liver, entirely located on the right, between the last rib and the 6^{e} intercostal space, is also accessible. The abomasum is also visible along the hypochondrial circle. The duodenum extends the abomasum posteriorly, and the pancreas lies deep beneath the caudate lobe of the liver.

[1] https://www.youtube.com/watch?v=dZyeKmytFeA

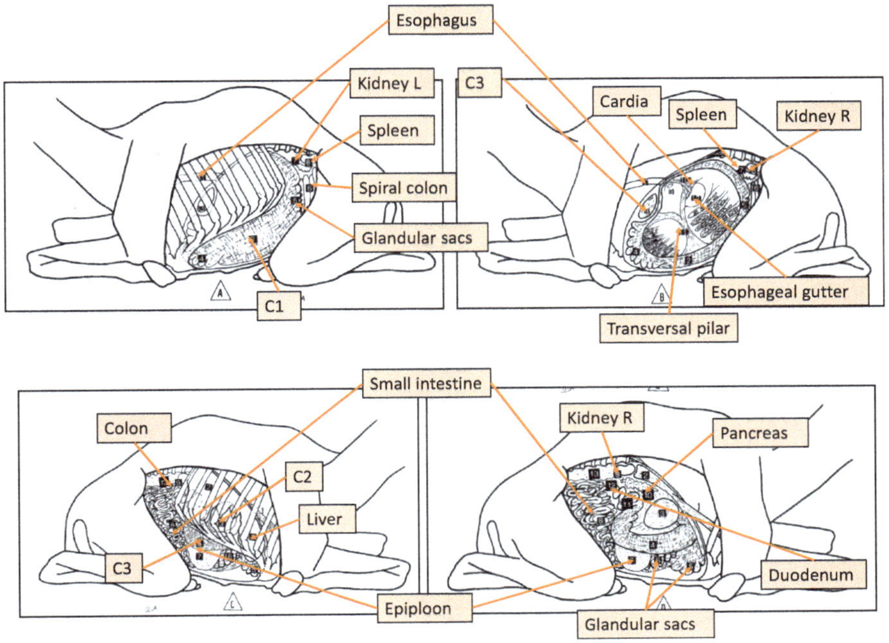

Fig. 2.19 Abdominal topography in the bearded dromedary, seen from the left lateral side (upper figures) and right lateral side (lower figures) before and after flap resection. (After Oushine 1989)

The right kidney projects into the hollow of the flank against the first two lumbar vertebrae.

References

Oushine A (1989) Étude de la topographie des viscères abdominaux chez le dromadaire (*Camelus dromedarius*) en décubitus sternal. Rev Elev Med Vet Pays Trop 42(1):73–78

Senault R (2013) Biomécanique des quadrupèdes. Entre diagonalisation et latéralisation. Mémoire de master, Université de Montpellier, France, 199 p

Physiology of Large Camelids: Life Cycle, Adaption to Ecosystems, and Reproduction

3

Abstract

The life cycle of large camelids is characterized by late puberty, long gestation, low fecundity rate, high mortality of the young, and long intercalving interval but compensated by high longevity (long life expectancy). Such peculiarities lead to low numerical productivity of camel herds. The large camelids are also known for their adaptation to desert ecosystem, i.e., resistance to extremely hot (dromedary) or cold (Bactrian) weather, resistance to drought and scarce water resources, ability to support large intervals between watering and rapid rehydration, and ability to valorize low-nutritive plants, thanks to the specific physiology of digestion (buffer power of abundant saliva, long stay in stomach, slow intestinal transit, urea cycle). The sexual cycle of large camelids is characterized by the lack of visible heat, ovulation provoked by mating, and seasonal pattern of testosterone secretion in male leading to rutting period during winter. Despite a comparable udder shape to cow, the internal cistern is small, leading to small part of cisternal milk (<15%) and a milk flow often marked by bimodal ejection. Compared to the cow, lactation curve has less marked peak, higher persistence rate, and longer lactation in high-yield dairy camels. Their feeding behavior is marked by ambulatory grazing, use of large floristic facies, and relatively low intake.

Keywords

Physiology · Reproductive cycle · Adaptation · Lactation · Digestion

3.1 The Life Cycle of Large Camelids

The large camelids (dromedary and Bactrian) are regarded as low prolific species. In a productive life and despite a certain longevity (20–30 years), the female produces only 6–8 camels at best. Some are considering such low performance in terms of

B. Faye et al., *Large Camel Farming*, https://doi.org/10.1007/978-94-024-2237-5_3

Table 3.1 Some reproductive parameters in large camelids (range and mean)

Parameters	Interval	Average
Age at puberty	30–48 months	36 months
Age at first birth	42–72 months	54 months
Gestation period	370–390 days[a]	380 days
Twinning rate		0.4%
Calving interval	15[b]–36 months[c]	24 months
Number of births per career	3–8	6
Annual fertility rate of the herd	35–40%	35%

[a]Longer gestation in the Bactrian camel (two humps), up to 410 days
[b]In camels not used for milk production and having calved at the beginning of the breeding season
[c]Especially in the case of abortion

numerical productivity as an adaptation to difficult conditions. In fact, in environments where food availability is low, such as in desert or semi-desert environments, the birth of a young camel is always a gamble, firstly for the newborn, which is always fragile (the mortality rate in extensive environments can reach 20%), but also for the mother, who is weakened by giving birth and who draws on her body reserves (especially her hump) to suckle her calf and ensure it grows sufficiently to overcome its fragility. In more intensive environments with a richer and more regular diet and more favorable sanitary control conditions, an improvement is palpable, both in the survival rate of the young and in the fertility rate of breeding females, contributing to a better numerical productivity. But this has a relatively little impact on the life cycle, which remains more or less the same. Given the known reproductive parameters of large camelids (Table 3.1), their life cycle can be described as follows:

- The birth of the baby camel occurs after gestation during the winter season (in the broad sense: between November and March in the northern hemisphere), because of the seasonality of the reproduction periods linked to the rutting period in the male, which takes place in the winter period.
- The growth of the baby camel, if it survives, is ensured by the milk production of the mother at least in the first months of its life: in extensive systems, weaning is late, rarely before 6 months.
- Puberty occurs on average around 3–4 years of age, with the first births occurring around 4–5 years of age. The age of puberty depends on the weight of the animal (on average around 70%–80% of the adult weight) and therefore on the prepubertal diet.
- The gestation period is on average 390 days, or nearly 13 months, and varies according to the sex of the camel, its weight, and the season of birth. It is longer in the Bactrian.

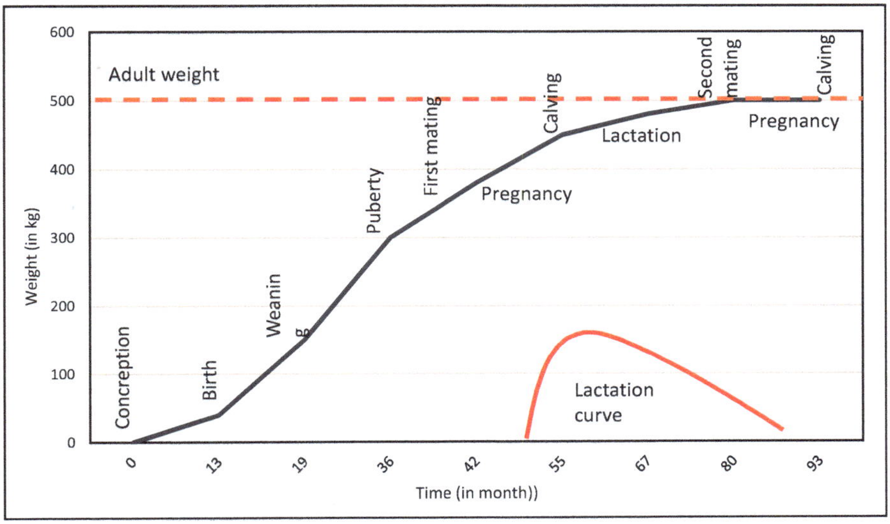

Fig. 3.1 Schematic representation of the first years of the life cycle of a dairy camel

- Lactation can last from 6 to 18 months, with an average of 12 months. As milk production stops rapidly after fecundation, there is an incompatibility between gestation and lactation. This results in an average calving interval of about 2 years, with 1 year devoted to producing milk and 1 year devoted to producing a fetus.
- Adult weight is reached around the age of 7.
- The productive life of the camel is in the order of 10–16 years, although animals reaching 30 years of age can still be found, provided there are good feeding and sanitary conditions.
- The culling of females comes late (over 20 years old), and in some countries, it is forbidden (Saudi Arabia, e.g., except in cases of force majeure).
- Males not intended for reproduction are slaughtered either before the age of 2 (Gulf countries), or when they reach 250 kg (Tunisia), or after puberty (Sahelian countries).
- Males intended for breeding reach full puberty at about 4 or 5 years of age, although they may show sexual behavior before that age. However, their first breeding is rarely done before 5 or 6 years. They are used as reproducers for 5–10 years depending on the turnover established by the breeder.

In the end, reproductive performance appears to be poor, and improving it remains a constant demand from breeders. The methods of improvement concern the feeding of the young to accelerate the different phases of the beginning of the life cycle (age at puberty, age at breeding, age at first parturition), the sanitary control of the newborns (increase in the survival rate), and the reproductive capacities of the females (improvement in the fertility rate).

Figure 3.1 summarizes the life cycle of a camel up to the first gestation/lactation cycle.

3.2 Mechanisms of Adaptation

Naturalists have always wondered about the dromedary's physiological ability to adapt to desert life, but it was not until the 1950s that fundamental work on dromedary physiology led to a better understanding of the adaptive mechanisms to arid ecosystems. Indeed, most mammals living in deserts adapt to heat and drought by burying themselves in the ground during the hot hours, a behavior that is impossible for an animal having the size of a large camelid. It was therefore necessary for the dromedary and the Bactrian to develop other arguments to adapt to the conditions of its original ecosystem.

3.2.1 Adaptation to Heat

The fat reserve of the hump plays an essential role in thermoregulation because of its concentration on the animal's back. Indeed, such a configuration limits the dispersion of subcutaneous fat under the skin and thus facilitates the dissipation of heat through the skin. Dehydration, which for most species is accompanied by an increase in blood viscosity leading to a rise in internal temperature that is often fatal (heat stroke), does not affect the dromedary's blood, which remains fluid, enabling it to maintain its heat transfer function from the periphery (cooler due to evaporation) to the heart.

The dromedary also has the capacity to vary its internal temperature according to the external heat: thus, when the ambient temperature decreases, especially during the night, the internal temperature of the dromedary can go down to 34 °C. During the hottest hours, the rectal temperature can reach 42 °C without being considered as a fever. Such changes in body temperature are fatal for most mammals. This mechanism, by decreasing the temperature differences with the outside world, represents a powerful instrument of adaptation to the temperature variations between day and night. Thus, an increase of 6 °C of the body temperature would save 5 liters of water.

Fig. 3.2 During the hot hours, the dromedary faces the sun to reduce the skin surface exposed to solar radiation (Mauritania)

The anatomical features mentioned above also indicate an adaptation to heat: when the ground is hot and the animal is standing, its long limbs isolate the body mass from the high ground temperature; when the ground is colder, even when the dromedary is standing in a stooped position, the sternal pad keeps the abdomen slightly above the ground, allowing air to circulate and dissipate the heat. In addition, during hot weather, the camel will prefer to stand facing the sun in order to expose the smallest possible area of its body to maximum solar radiation (Fig. 3.2). When grazing, it prefers to graze on woody forage rather than grasses in order to take advantage of the shade of the trees while feeding, reserving the steppe grassland areas for the beginning and end of the day.

The increase in external temperature normally leads to an increase in the general metabolism by overflow of the thermoregulatory mechanisms for which the thyroid gland plays a central role, because the activation of the general metabolism is under its dependence. As thyroid activity increases with external heat, metabolic heat production increases, maintaining a vicious circle leading to "heat stroke," which can be fatal. Conversely, in the dromedary, the increase in internal temperature is associated with a general decrease in metabolism, which is reflected in a decrease in oxygen consumption. In other words, it is able to depress internal heat production during the hottest part of the day. Heat and dehydration have been shown to increase bromide secretion from the thyroid gland, depressing thyroid activity and thus slowing the overall metabolism (Etzion et al. 1987).

Large camelids also have a winter fleece (especially Bactrians) that falls out by itself in summer (Fig. 3.3). Such fleece, abundant during the cold period of the year,

Fig. 3.3 Molting of a Bactrian camel at the onset of heat (Kazakhstan)

is an excellent insulator that creates a temperature gradient from the skin surface to the end of the wool fibers, thus reducing skin water loss. Bactrian fleece, in particular, is known to have superior insulating and anti-UV properties compared to other wool species (Hasi et al. 2020), which led the institutions of space conquest in the Soviet era to equip cosmonauts with camel wool-based underwear.

The color of the coat also plays a role in heat adaptation processes, with light colors (white to fawn fleece), which better reflect the sun's rays, being predominant in camel populations (Fig. 3.4). Finally, it can be noted that with its thick, protective skin, equipped with a system of sweat glands that only becomes active when the body temperature exceeds 42 °C, the camel has a large heat storage capacity.

3.2.2 Adaptation to Drought

The mechanisms of adaptation to heat involve physiological procedures that also contribute to saving water, as outlined in the previous chapter. However, it is in extreme situations, especially during severe dehydration, that the camel shows its exceptional qualities of adaptation to water deprivation. The ability of camelids to go without drinking for several days, or even weeks, if the food ration is palatable (rich in water and salt), is sufficiently well known to avoid having to come back to it. But, on the contrary, the dromedary is also able to ingest a large quantity of water in a minimum of time after a long session of water deprivation, and this, without

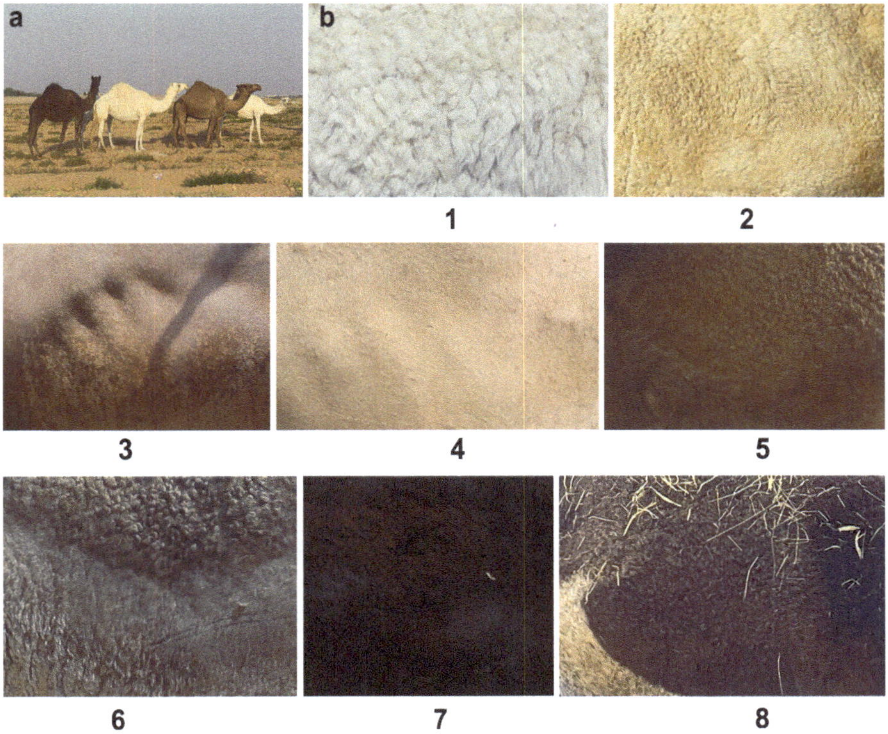

Fig. 3.4 (**a**) Coat color variability in large camelids (Saudi Arabia); (**b**) range of coat colors in dromedary: (1**b**) white, (2**c**) yellow, (3**d**) red, (4**e**) light brown, (5**f**) dark brown, (6**g**) gray, (7**h**) black, (8**i**) black piebald. (After Faye et al. 2011)

disturbing its metabolism. The ingested water is absorbed by the salivary glands, stomach, and intestine before reaching the intercellular spaces and not, as an ignorant public believes, the hump, which is not a water storage organ. The dromedary is in fact capable of losing a third of its weight in water without endangering its life and of quickly recovering its initial weight after watering. Such a loss of water, as well as such a rapid rehydration, is generally fatal in most species. Indeed, rapid and abundant watering causes an uncontrolled influx of water into the bloodstream, which leads to a dilution of the extracellular milieu (hypotonicity). To compensate for the difference between this hypotonic milieu and the hypertonic internal cytoplasm of the red blood cells, the latter swell with water until they burst (hemolysis). However, the dromedary has red blood cells that are extremely resistant to hypotonicity due to the composition of its cell wall (Al-Qarawi and Mousa 2004). Overall, after rehydration, the return to a cellular and metabolic state prior to water deprivation takes about 4 hours for blood cells and kidney functions.

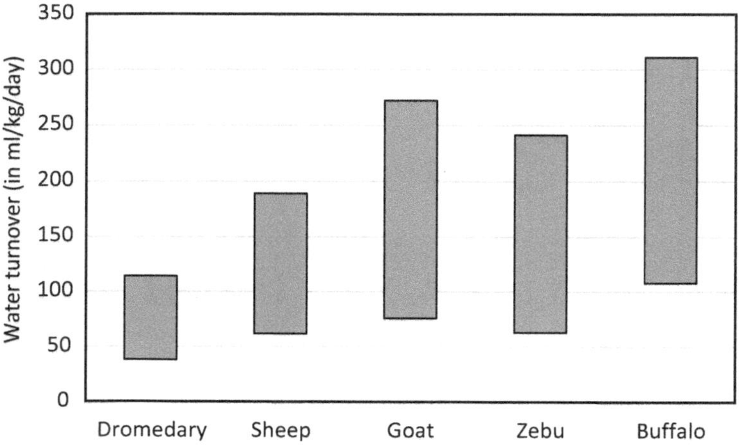

Fig. 3.5 Comparison of water turnover between different domestic species

Water turnover in dromedaries is particularly low compared to other species (Fig. 3.5), although the percentage of body water is higher: 72 vs. 60–65% in other domestic herbivore species.

Water absorption begins with the salivary glands. A normally hydrated camel produces a large amount of saliva (about 20 l/d per gland!), but in the dehydrated animal, saliva production can drop to 1 l/d without affecting intake.

However, the main storage organ for ingested water is the stomach. As long as possible, a thirsty camel will maintain a high proportion of water, especially in the C1 compartment (rumen), which ensures a smooth passage of ingesta into the other gastric compartments and into the intestine, thus maintaining its appetite. Massive water reabsorption takes place at the terminal level of the intestines, resulting in the excretion of particularly dry droppings. Indeed, the percentage of dry matter in dromedary excreta is about 50%, compared to 30% for small ruminants and 15% for cattle. In dehydrated dromedaries, the fecal dry matter rate increases slightly (55%), whereas it does not change in cattle. In the end, the dromedary would lose seven times less water through the fecal route than cattle.

In the mechanisms of adaptation to drought, urinary excretion is obviously essential. In this respect, the kidney is a central organ in water regulation, contributing, in case of dehydration, to a strong reabsorption of water and electrolytes and to the excretion of a concentrated urine. Urine volumes of dehydrated camels and sheep were compared: the volume was 0.1% of the camel's weight compared with 2% in sheep. Dehydrated camel urine is twice as concentrated as sea water. This ability to eliminate highly concentrated urine helps explain the dromedary's high salt tolerance and its ability to cope with brackish water. The wild Tartarian camel has taken this ability to the extreme, so much so that it could survive drinking only seawater. Thanks to this mechanism, at a temperature of 42 °C, the dromedary loses about 20 g of water/kg of body weight (BW)/day, compared to more than 40 g in sheep.

Fig. 3.6 Desert environments are characterized by thermal amplitudes, drought, and scarcity of food resources: camel in the Khowaar Desert (Saudi Arabia)

Globally, large camelids implement two complementary mechanisms to fight against dehydration: a minimization of water losses and an optimization of water resources (Bengoumi and Faye 2002).

3.2.3 Adaptation to Malnutrition

While arid environments are hot and dry ecosystems, they are also characterized by limited food resources, wide spatial dispersion, and high seasonal and even interannual variability (Fig. 3.6). In such a context, large camelids only survive because of their ability to manage the low density and low nutrient value of desert vegetation. We will see in Chap. 4 ("Physiology of Digestion") how this can explain the capacity of large camelids to digest and valorize poor forage better than other domestic herbivores. Adaptation to energy undernutrition through the management of fat reserves concentrated in the hump (see Box 6.1 on body condition score) and adaptation to protein undernutrition through the efficiency of the nitrogen cycle (see Sect. 4.2.1) are indeed characteristics of the species. On the other hand, overly rich food rations, especially in more favorable latitudes (western countries), can be the cause of severe metabolic disorders.

Adaptation to undernutrition also concerns major and minor minerals which, although widely present in mineral-dominated ecosystems, can lead to situations of deficiency or toxicity (see Sect. 5.4.1.4).

3.3 The Physiological Basis of Reproduction

As this book is general in nature, we will limit ourselves here to the main points that need to be known in order to ensure one of the important functions of livestock farming. For more details, please refer to the remarkable work by Tibary and Anouassi (1997) cited in the references.

3.3.1 The Sexual Cycle in Camels

The camel, unlike the cow, does not express very visible "heat." However, it is not without an estrous cycle, although for large camelids, the term appears debatable. Indeed, the main characteristic of the reproductive physiology of the female dromedary or Bactrian is that ovulation is caused by mating. Consequently, the complete cycle only takes place in the presence of the reproductive male and in case of effective mating. In the absence of the male, the cycle is limited to a succession of "follicular waves" with a periodicity that varies from 3 to 4 weeks. Once the follicle has matured, it will only release the ovule if there is a mating. If there is no fecundation, the follicle regresses, allowing the next one to develop. If the ovule is fecunded, the follicle is maintained and becomes a gestational follicle or "corpus luteum."

The "follicular waves" have several phases, related to hormonal changes, which are described as follows (Fig. 3.7):

- A growth phase lasting 2–6 days, during which several follicles increase in size and emerge on the surface of the ovary. Their sizes reach about 2–3 mm.

Fig. 3.7 Dromedary follicles at different sizes and degree of maturation. (Source: Tibary and Anouassi 1997)

- A dominance phase, during which only one or two of the growing follicles reach a size of about 8 mm in diameter. This phase also lasts from 4 to 6 days.
- A maturation phase which concerns the dominant follicle (more rarely two in number), which lasts on average 13 days (between 5 and 19 days depending on observations) and results in a growth in diameter, then a relative stagnation in size when it reaches the pre-ovulation stage. This size can reach 25 mm. Ovulation occurs at the time of mating. In this case, the follicle evolves toward the constitution of a corpus luteum whose characteristic in large camelids is to develop slowly but also to disappear quickly in case of nonimplantation of the embryo.
- A regression phase which occurs in the absence of mating and lasts from 4 to 18 days. The follicle becomes atretic or cystic depending on the case. The first case is a natural degeneration lasting from 4 to 18 days. The second case corresponds to the evolution toward an anovulatory cyst, sometimes hemorrhagic, of slow regression (up to 45 days) and which delays the emergence of a new cycle.

The question for the breeder is to detect the period during which the female accepts mating. As almost half of the camels show no or weak signs of heat (mounting of a female by another female), and as these signs may occur outside the preovulatory phase, it seems difficult to rely on these behavioral changes to detect the appropriate period. This is why in most cases, the male is put in the presence of the females during the breeding period, as only he can detect the estrous phase.

3.3.2 The Sexual Cycle in Males

The male dromedary or Bactrian is characterized by a seasonal sexual cycle. Indeed, a rutting period occurs during the winter months (on average between November and March in the northern hemisphere) probably in relation to the temperature and length of the day. The sexual activity of the male is dependent on testosterone, which, in large camelids, increases by a factor of 10 or even 15 between the periods of sexual rest and the rutting period. Indeed, serum testosterone levels rise from a baseline of 2 ng/mL to a maximum of 35 ng/mL in the depths of winter (fig. 3.8).

In concrete terms, such a cycle results in morphological changes in the testicles, which increase in size during the rutting period (Fig. 3.9) and, above all, in behavioral changes.

During the rutting period, the breeding male shows the following behavioral characteristics:

- Aggressiveness: during the rut, the male becomes aggressive and can attack other males in real fights, but also people, and the injuries it imposes (especially bites) can be dangerous. The rutting male must therefore be handled with care even if, during periods of sexual rest, it appears very gentle and friendly with people.

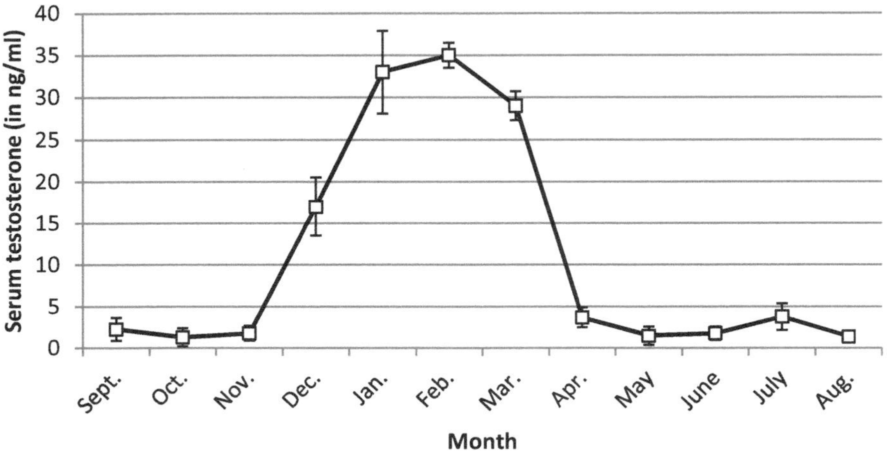

Fig. 3.8 Changes in blood testosterone levels in male dromedaries over the year. (Source: Faye and Bengoumi 2018)

Because of this aggressiveness, it loses its appetite and loses weight. He can lose up to 35% of its weight. Sometimes its intestinal transit is accelerated, and it presents diarrhea. It is based on this behavior that camel wrestling festivals are organized in Turkey. It is therefore important to avoid having several breeding males in the same herd during the rutting season, especially as there is competition for the domination of the female herd.

- Externalization of the soft palate: during the rut, the male externalizes the soft part of its soft palate (the wall separating the oral cavity from the nasal cavity), an anatomical feature of large camelids known as the *dulla* (Fig. 3.10). This behavior may be repeated regularly throughout the rutting season. Note that this externalization (which the uninformed confuse with the tongue) occurs most often on the right side of the mouth, making an unpleasant gurgling sound. The phenomenon is accompanied by the emission of a foamy drool.
- Teeth grinding: the rutting period is also the time of the appearance of teeth grinding by lateral friction of the molars emitting an unpleasant metallic noise
- Occipital gland excretion: the male has two glands at the back of the head (known as occipital glands), which secrete a dense, opaque, almost blackish-brown fluid, particularly abundant during the rut (Fig. 3.11). This foul-smelling fluid contains androgenic steroids and pheromones with which the animal marks its territory (when outside) by rubbing itself on trees or enclosure features.
- Urinary marking: marking of the territory is not only done with the liquid from the occipital glands but also with the urine that the animal scatters by taking a particular posture consisting of spreading the hind limbs and spraying its urine by vertical movements of its tail around and over it, so that after a few weeks, the hindquarters are covered with a blackish crust, a mixture of urine and earth or sand (Fig. 3.12).

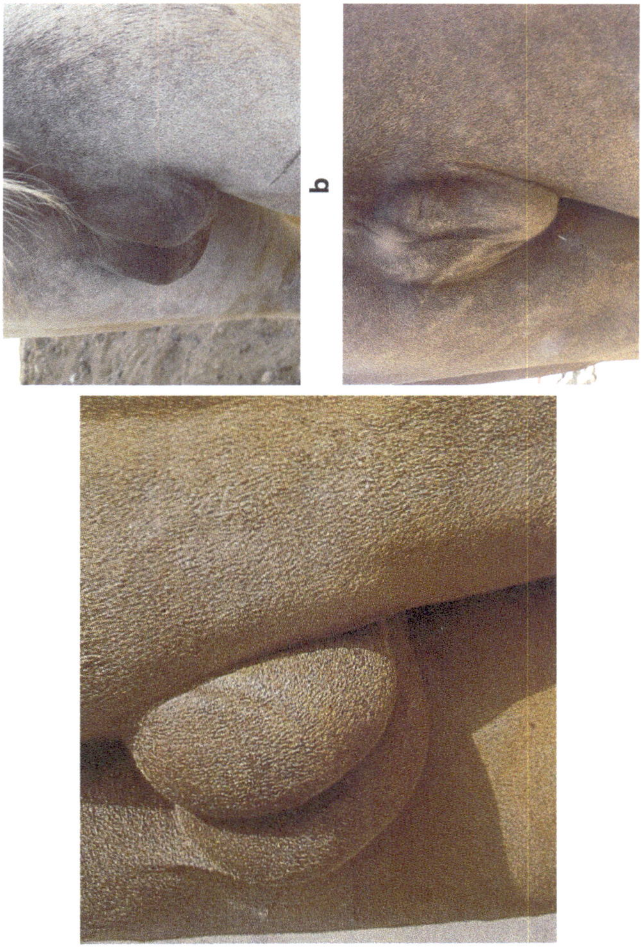

Fig. 3.9 Changes in testicular morphology; (**a**) mating period; (**b**) intermediate period; (**c**) sexual resting period

Fig. 3.10 Externalization of the soft palate in a male dromedary in rut

Fig. 3.11 Occipital gland excretion at the back of the head in a male in rut

Fig. 3.12 Urine marking posture of males during the rutting period. (Source: Tibary and Anouassi 1997)

- Pre-mating behavior**:** toward females in the period of acceptance of mating, the male pursues them, sniffs their flank or their perineum, and constantly defends his "harem" against other males, or even people present in the herd. He also frequently assumes a specific posture with his head and lips called *flehmen* (Fig. 3.13). He then forces the female to crouch for mating before crouching on the back of the hump to mate.

3.3.3 Mating

Mating in large camelids takes place in the sitting position for the female (Fig. 3.14). The characteristic feature of this large herbivore species is the duration of copulation, which can vary from 5 to 25 minutes depending on the mating season and the age of the mating animal. Because ejaculation occurs throughout copulation, the success rate increases with the duration of mating. In some cases (young inexperienced males, potentially aggressive males), the breeder may have to assist the male by facilitating the intromission of the penis, thus avoiding sometimes dangerous interactions and the introduction of sand into the female's genitals.

Information on management practices to improve the numerical productivity of a camel herd can be found in Chap. 5 "Reproductive Management."

Fig. 3.13 Characteristic *flehmen* posture in the dromedary male before mating

Fig. 3.14 Assisted mating in dromedaries

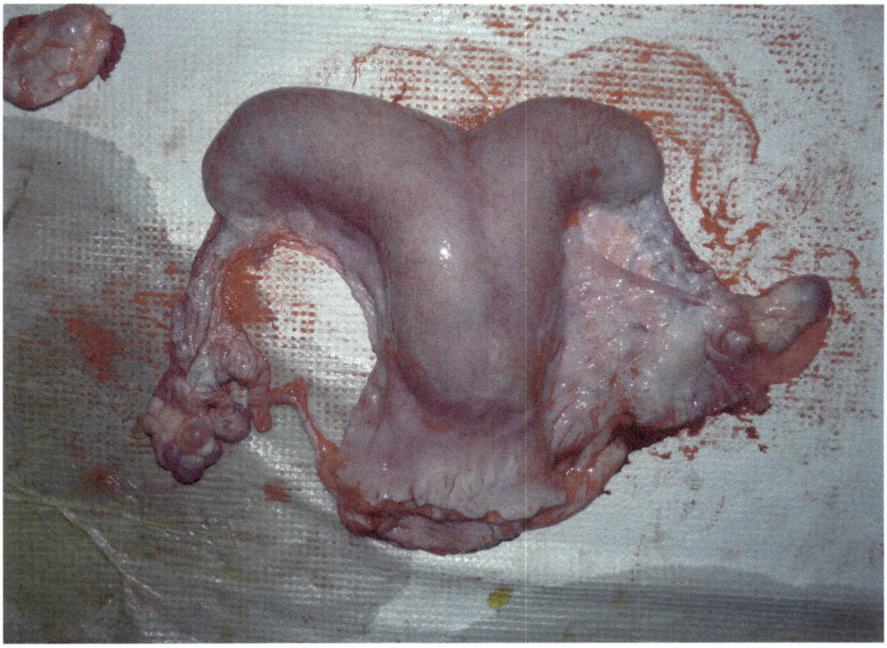

Fig. 3.15 Camel uterus. The left horn appears more developed than the right horn; it is the place of gestation

3.3.4 Gestation

Gestation in camels is long (average of almost 13 months) and almost always occurs in the left uterine horn, the uterus being bifid (Fig. 3.15).

The embryo is implanted in the uterus from day 6[e] after ovulation and attaches itself to the uterus by day 20[e]. Although the embryo may appear in the right horn of the uterus, in almost all cases, it migrates to the left horn, probably because the left horn is more developed. However, the mechanism of this migration is unknown. Even if, sometimes, multiple ovulations can occur naturally, the birth of twins is very rare: between 0.1 and 0.4%. Early embryonic mortality is also quite frequent, as well as abortions, which can affect 10% of pregnancies in some herds. The placenta of large camelids is of the diffuse epitheliochorial type, similar to that of the horse and therefore very different from that of the cow. As a result, retained placenta is rarely observed in camels after parturition. In all camelids, there is an extrafetal membrane called the epithelion, which completely covers the fetus and avoids direct contact with the amniotic fluid, unlike other species. This membrane acts as a lubricant during parturition and protects against dehydration.

3.3.5 Parturition (Box 3.1)

The first signs of childbirth (Box 3.1) can be seen a few weeks before. The first sign is the development of the mammary gland (about 2 weeks before), which is particularly visible in primiparous female. Then, about 15 days before parturition, there is the beginning of relaxation of the sacro-sciatic ligament, which is located on each side of the pelvis, and an edema of the vulva (Fig. 3.16), which appears a few days before.

The actual delivery consists of three phases:

- A preparatory stage that can last from 3 to 48 hours, during which the mother-to-be shows anxiety, does not stop moving, isolates herself from the herd, and, in some cases, reveals the presence of her water bag (Fig. 3.17).
- A stage of expulsion of the fetus, which starts with the rupture of the water bag and has a variable duration, on average between 5 and 45 minutes depending on the strength of the contractions, the age of the mother, and the weight of the camel. In most cases, the fetus comes out head-first resting on the front legs. The female gives birth lying on her side but sometimes standing. Dystocia (difficulty in delivery) is uncommon in this species (Fig. 3.18);
- A stage of expulsion of the placenta, which is quite rapid in large camelids (about 40 minutes). The placenta is quite heavy, up to 10 kg in the heaviest newborns.

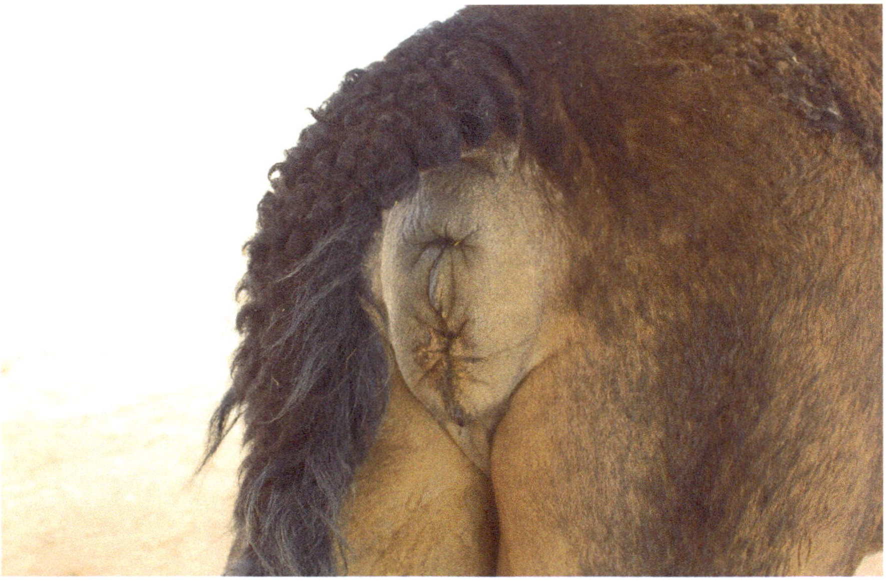

Fig. 3.16 Edema of the vulva in an early parturition camel

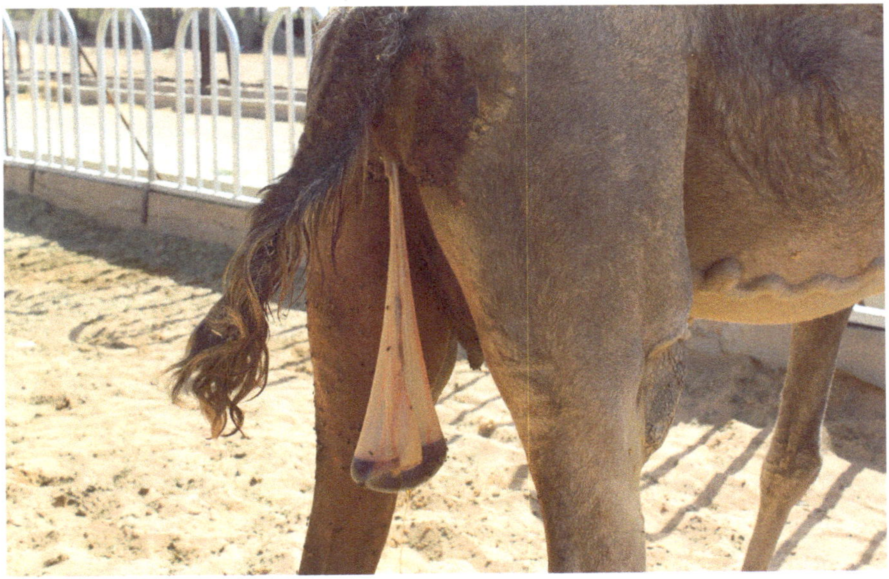

Fig. 3.17 Appearance of the water bag, signaling the start of fetal expulsion

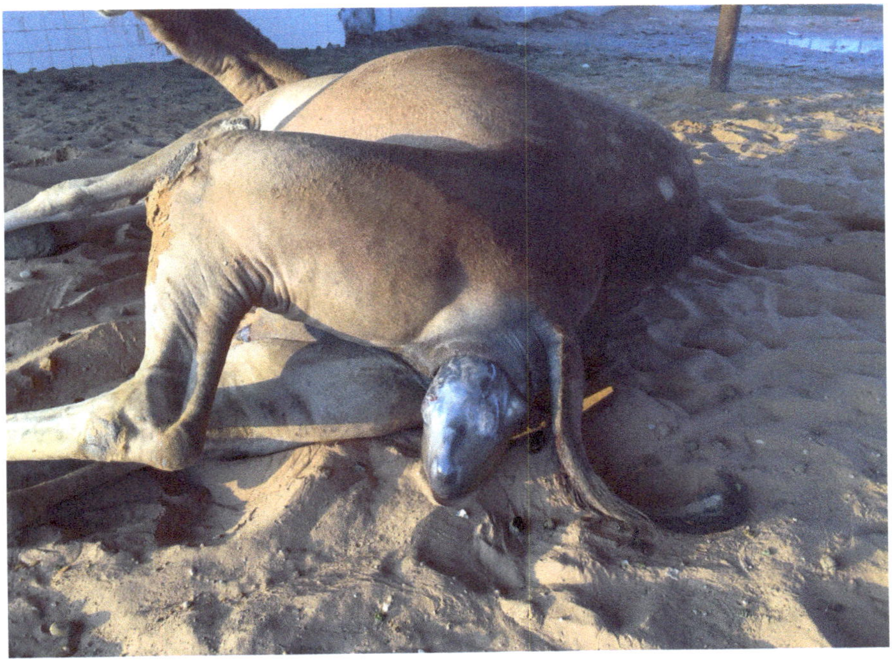

Fig. 3.18 Fetal expulsion phase of camel delivery

Fig. 3.19 Baby camel just after calving still wrapped in the epithelion

After birth, the camel calf can sit up within 5–15 minutes (Fig. 3.19), stand up quickly, and be able to suckle within the first hour.

The maternal instinct is expressed from birth by the recognition of the calf by its smell (Fig. 3.20). During delivery, the hormonally induced rise in milk production can lead to a natural ejection of milk.

Vulvar and uterine involution after delivery occurs within a few days for the vulva and a few weeks for the uterus (about 3 weeks). Lochia, postpartum vulval discharge, is low in large camelids.

The first heat return after birth occurs at a variable interval. Cyclicity can resume immediately after birth – especially if the calf does not survive – and under good feeding conditions.

But, on average, postpartum anestrus lasts about 30 days, and mating can occur as early as 45^{e} days postpartum. However, the ovary often remains inactive during part of the lactation: this is the lactation anestrus. Additionally, if breeding is not incompatible with lactation, milk production drops and then stops within a few weeks after the start of gestation (see Chap. 5 "Reproductive Management") as observed by Nagy et al. (2015).

Fig. 3.20 Recognition of the baby camel by its smell by the mother a few minutes after giving birth

Box 3.1 Stages of Eutocic Delivery in Camels (Fig. 3.21)

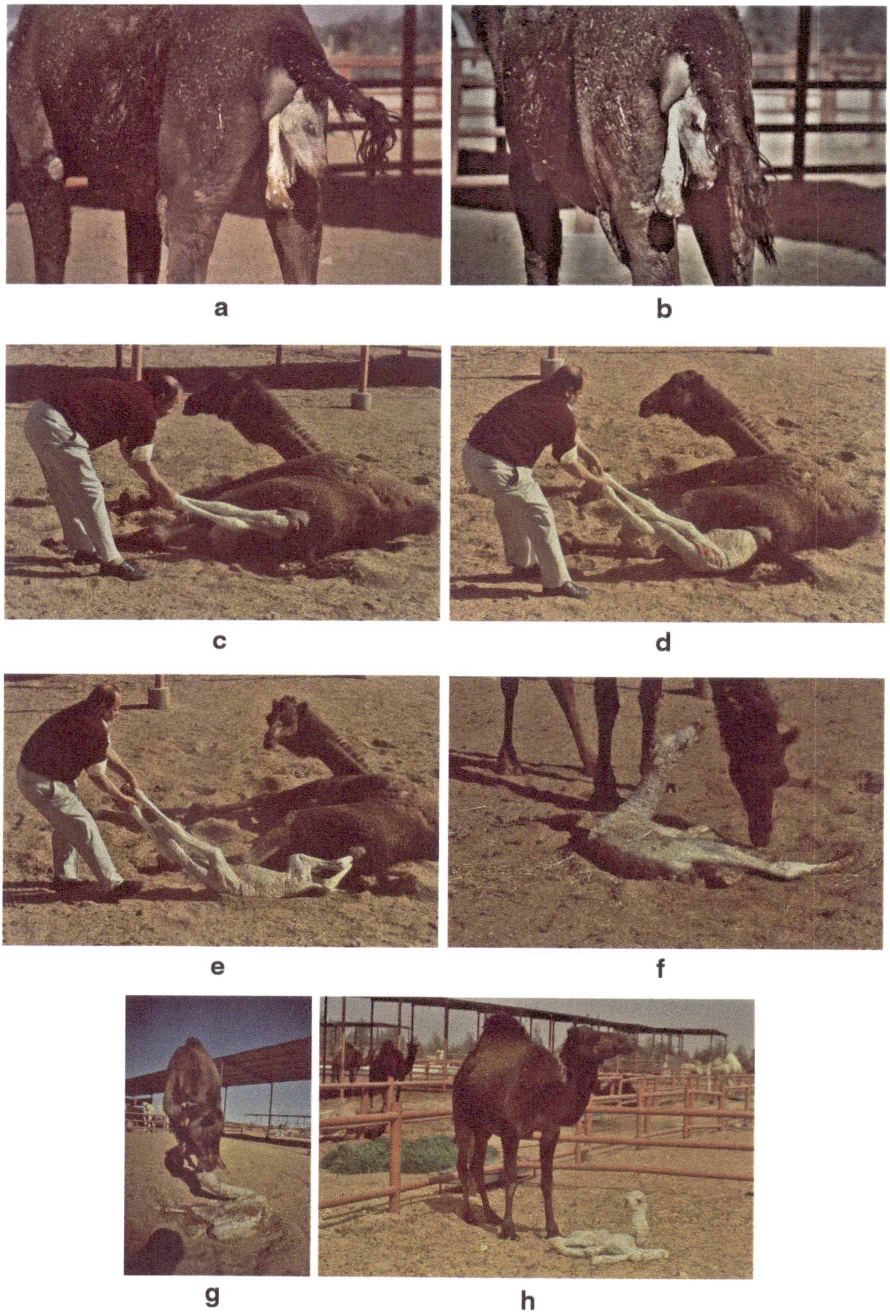

Fig. 3.21 (**a** and **b**) The newborn presents in the "normal" position (head and forelimbs first). (**c**–**e**). The practitioner only facilitates expulsion, with the mother lying on her side (usually the female remains standing) by accompanying the contractions. (**f**–**h**). When the calf comes out, the mother gets up and recognizes her calf by the smell

References

Al-Qarawi AA, Mousa HM (2004) Lipid concentrations in erythrocyte membranes in normal, starved, dehydrated and rehydrated camels (Camelus dromedarius), and in normal sheep (Ovis aries) and goats (Capra hircus). J Arid Environ 59:675–683. [on line]. https://doi.org/10.1016/j.jaridenv.2004.02.004

Bengoumi M, Faye B (2002) Adaptation du dromadaire à la déshydratation. Revue Sécheresse 13: 121–129

Etzion Z, Alfassi Z, Lavi N, Yagil R (1987) Halide concentration in camel plasma in various state of dehydration. Biol Trace Elem Res 12:411–418

Faye B, Abdallah H, Almathen F, Harzallah B, Al-Mutairi S (2011) Camel biodiversity. Camel phenotypes in the Kingdom of Saudi Arabia. Camel Breeding, Protection and Improvement Center, project UTF/SAU/021/SAU, FAO publ., Riyadh (Saudi Arabia), 78 p

Faye B, Bengoumi M (2018) Camel clinical biochemistry and hematology. Springer, New York, 346 p

Hasi S, Amu G, Zhang W (2020) Camel hair structure, properties, and commercial products. In: AlHaj O, Faye B, Agrawal RD (eds) Handbook of research on health and environmental benefits of camel products. IGI Global, Hershey, pp 328–347

Nagy P, Faigl V, Reiczigel J, Juhasz J (2015) Effect of pregnancy and embryonic mortality on milk production in dromedary camels (*Camelus dromedarius*). J Dairy Sci 98:975–986. https://doi.org/10.3168/jds.2014-8546

Tibary A, Anouassi A (1997) Theriogenology in camelidae, Actes éditions IAV Hassan II. Rabat, Maroc, 489 p

4 The Physiological Basis of Large Camelid Lactation, Digestion, and Nutrition

Abstract

Despite the external appearance similar to that of a cow, the camel udder anatomy is different, marked by a strong teat shape variability and a low cisternal part leading to a more difficult milking, especially when machine milking is used. Three main milk flows are described, allowing the identification of camel types according to their milking ability. The lactation curve is similar to that of a cow but with less marked peak (rather a plateau), higher persistence rate, and usually longer lactation duration. The digestion in large camelids is characterized by a strong buffer power, thanks to abundant saliva, long stasis of solid particles, and slow intestinal transit, leading to higher digestibility of poor-nutritive forages. Thanks to its browsing behavior, the camels are using a wide variety of floristic floors from grass to trees and a large panel of plant species in the environment contributing to a lower pressure on the resources. In addition, the camel has lower water requirements (80 ml/kg LW/24 h, i.e., half needs of cow) and lower feed intake (around 2 kg DM/100 kg LW); thus it is considered an environmentally friendly animal.

Keywords

Milk flow excretion · Udder morphology · Feed digestion · Browsing behavior · Watering

4.1 Physiology of Lactation

Camel is a dairy animal used for providing milk to humans since ancient time. However, the recent intensification of the milking process has boosted studies regarding the physiology of lactation, which appears different than for cow, despite a comparable anatomy of the udder.

B. Faye et al., *Large Camel Farming*, https://doi.org/10.1007/978-94-024-2237-5_4

4.1.1 Udder Morphology

Like the cow, the camel has an udder consisting of four quarters, but not separated by a furrow. The mammary organ is implanted under the abdomen (inguinal udder), and the teats are generally thinner and shorter than those of the dairy cow. The general morphology of the udder is variable from breed to breed, but due to the lack of selection on this trait, there is also a high degree of variability from one individual to another (Fig. 4.1).

The internal anatomy of the udder is characterized in large camelids by the following:

- The presence of several teat ducts (two or three), which has implications for local antibiotic treatment in cases of mastitis.
- A poorly developed cistern, in contrast to the lactogenic cells (alveolar acini), which leads to a lower reserve of "cisternal" milk (from 1% to 15% depending on the individual, with an average of 8%) compared with "alveolar" milk. The ratio of cisternal to alveolar milk is comparable to that of the buffalo but is very different from that of the dairy cow (about 35/65) and even more from that of the dairy goat (80/20).

Such a morphology has important consequences for the adaptation to the milking machine or the stimulation of the udder (see Chap. 5 "Milking management"). But above all, it helps to understand the dynamics of milk flow during milking. Indeed, the cisternal milk, stored between each milking in the cistern and the large milk ducts, is immediately available, as soon as the sphincter of the teat canal opens. The alveolar milk, stored in the acini, cannot flow on its own, and ejection requires contraction of the myoepithelial cells surrounding the acini. The opening of the sphincter, as well as the contraction of myoepithelial cells, is dependent on the hormone oxytocin.

4.1.2 Milk Flow

To measure the flow of milk, a device is used to record the volumes produced per time unit. This is an automatic recorder that is only operational when milking mechanically. It is placed between the udder and the teat cup (Fig. 4.2).

The data is recorded on an electronic medium linked to a software program that stores all the information in the form of tables and graphs. During milking, the milk flow consists of several phases:

- A stimulation phase, usually provided by the camel that starts the suckling, more rarely by massage of the teats by the milker. The stimulation triggers the secretion of oxytocin, which acts very quickly on the opening of the sphincter obstructing the teat canal, then on the acini cells, and consequently allows the milk ejection.

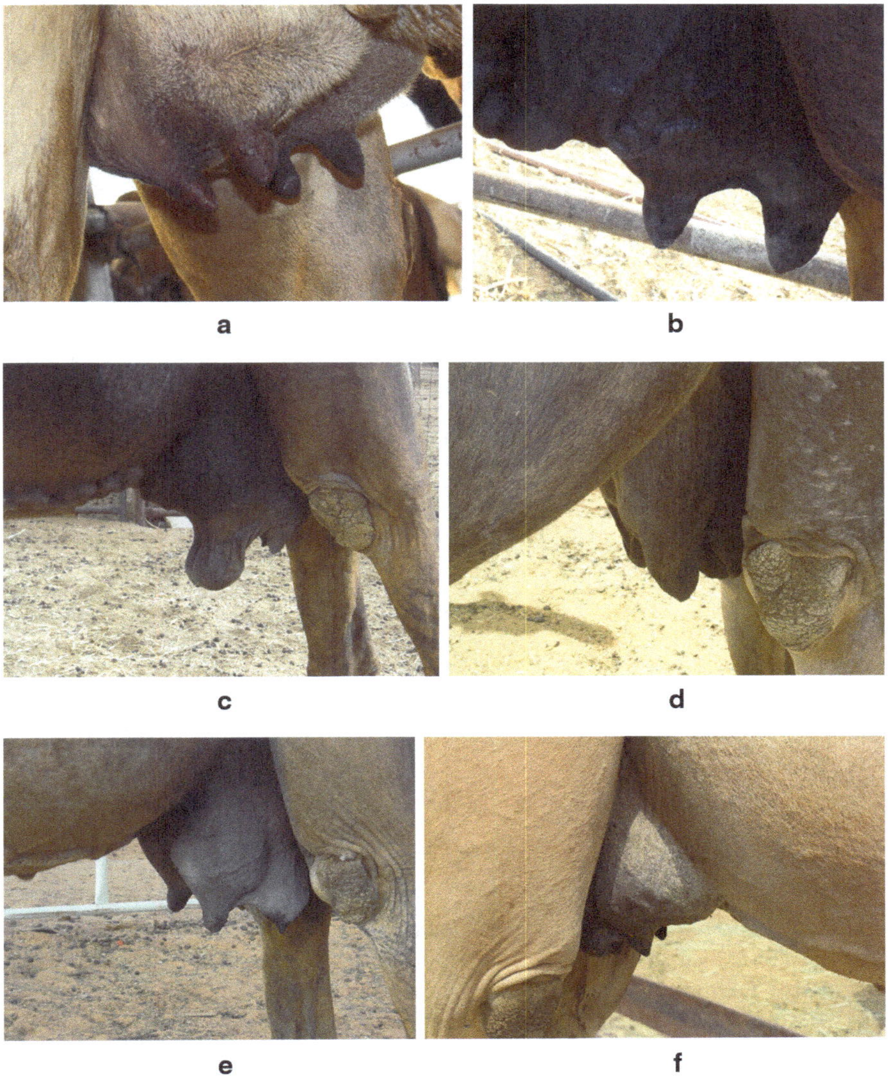

Fig. 4.1 Types of mammary morphology in the dairy camel. In general, there are six types of mammary morphology in camels: (**a**) cylindrical; (**b**) funnel-shaped; (**c**) bottle-shaped; (**d**) pendulous; (**e**) pyriform; (**f**) globular. (Ayadi et al. 2016)

- An ascending phase, which lasts a few seconds, during which the flow of milk gradually increases, reaching a maximum (rarely more than 2 l/min).
- A plateau phase, the duration of which varies according to the milk potential of the camel and during which the milk flow is more or less constant.

Fig. 4.2 Device for automatic recording of milk flow during milking. (The equipment is of the Lactocorder© type)

- A descending phase, also of variable duration, which ends with the cessation of lactation (Fig. 4.3).

There are several types of milk flow curves. At least three main types can be identified: (i) Fast milking, which is characterized by a very short rising phase with a peak of more than 2 or even 3/min, followed by a very short plateau and a progressive falling phase. This profile is characteristic of camels that are easy to milk, but not necessarily the highest producers. (ii) Bimodal milking, which is characterized by two peaks corresponding to the ejection of cisternal milk (first peak) then, in the second time, of alveolar milk. It could be related to insufficient stimulation and a delayed effect of the oxytocin hormone. (iii) Slow milking, where the plateau phase is replaced by a very progressive descending curve.

Chapter 5 ("Milking Management") explains how to interpret the readings of a milk recorder. Several studies have shown that milk production differs between the udder quarters. As in the case of the cow, the hindquarters are more productive than the forequarters: the hindquarters produce about 55–60% of the total milk.

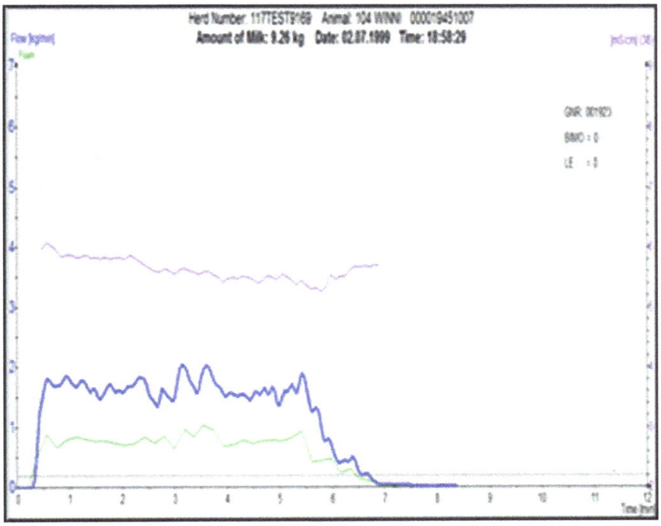

Fig. 4.3 Milk flow graph for a camel

4.1.3 The Lactation Curve

Milk production starts at calving, but the volume of milk produced is not constant during the lactation. This volume depends on many factors: genetics, feed, milking interval, quality of stimulation, weather conditions, stress and know-how of the milker or milking machine parameters, presence of diseases especially mastitis, and of course the physiological stage of the female, i.e., beginning, middle, or end of lactation. Overall, the lactation curve expresses the evolution of the production during the lactation. To describe this curve, several parameters are used (Musaad et al. 2013):

- The total duration of lactation, which varies from 6 to 18 months with an average of about 12 months and depends on the level of production of the camel and the quality of the diet. This should not be confused with the total duration of milk production, which is related to milking, the first few weeks being entirely devoted to suckling the camel.
- Initial production just after giving birth or at the start of milking.
- The duration of the growing production phase, which takes place over a period varying from 2 to 4 or 5 months depending on the individual.
- Production at the peak of lactation, which in camels is not very pronounced and is rather in the form of a plateau that can last several weeks. In these cases, the maximum production within 24 hours is taken into account.
- The time of maximum production (in days postpartum).
- The duration of the descending phase.

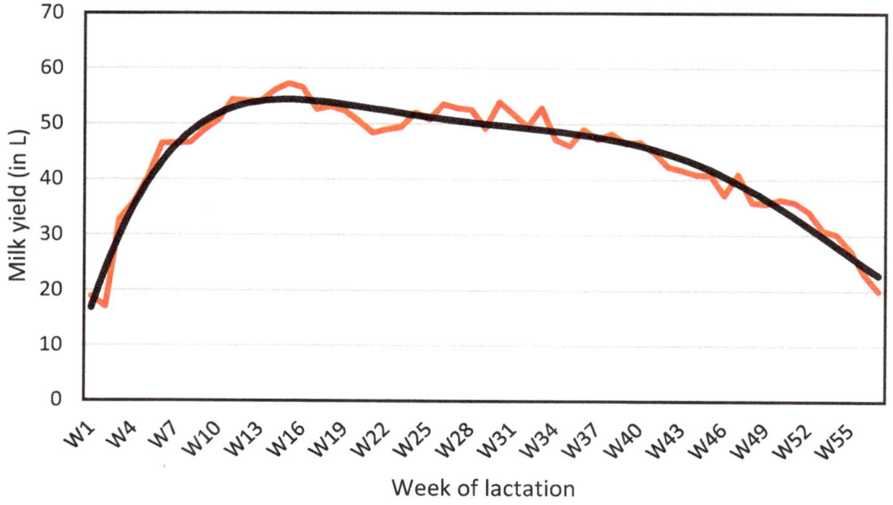

Fig. 4.4 Example of a real (red) and modeled (black) lactation curve for a camel of lactation rank 2 and calculated from weekly production. (Source: Aziz et al. 2016)

- The persistence rate, which reflects the slope of the decline in production and is estimated by calculating the decrease between two milk tests (usually 1 month) divided by the total duration of the downward phase. In camels, the persistence rate is higher than in dairy cows, i.e., the decrease in production is slower,
- The time of drying off, knowing that in camels, as production levels are relatively low, the cessation of production is mostly natural and does not require a water diet like in high-producing dairy cows.

The lactation curve (Fig. 4.4) varies according to the lactation rank of the mother (it is flatter in primiparous cows), the month of birth (it is shorter for late births in spring), and of course the level of production (the peak of lactation is less pronounced when the potential exceeds 3000 l of milk/year).

Models developed by several researchers allow the estimation of total expected production from the initial production, peak production, and persistency rate. The example in Fig. 4.5 shows two very different types of lactation curve for a low-producing camel (2.9 kg/day) and a high-producing camel (14.8 kg/day with a maximum of 30 kg).

4.2 The Physiological Basis of Digestion and Nutrition

Large camelids are pseudo-ruminants. This means that they have the physiology of ruminants, but they do not belong to the ruminant group from a taxonomic point of view (suborder Ruminantia). Dromedaries and Bactrians have physiological

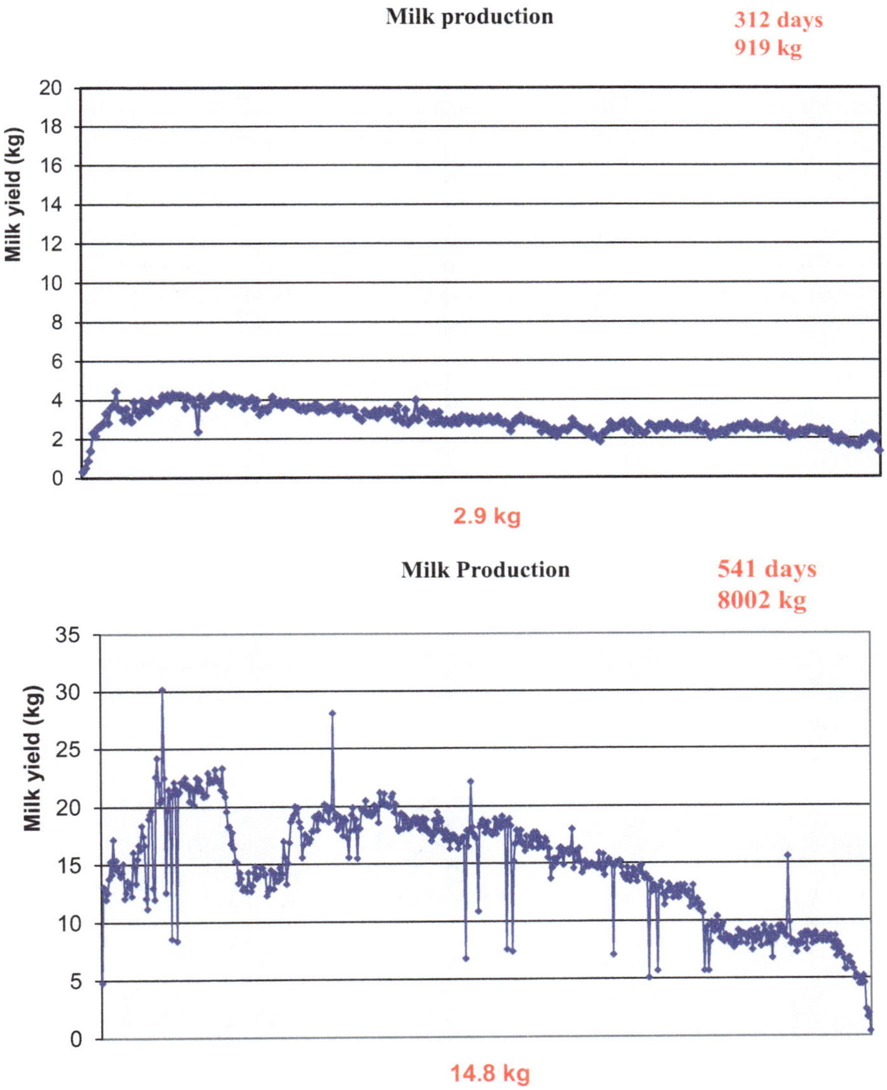

Fig. 4.5 Example of two extreme curves (low producer at the top and high producers at the bottom) for two camels from *Camelicious* Farm in Dubai. (Source: Nagy and Juhasz 2009)

peculiarities in the digestion of nutrients that impact their feeding behaviors and types of diet, even though, as herbivores, they are likely to use the same food resources as true ruminants.

4.2.1 Digestion in Large Camelids

Like all ruminant herbivores, large camelids are polygastrics. They have several stomachs with specific functions. However, the anatomy of the stomachs differs significantly. While ruminants have four stomachs, large camelids have only three compartments, named C1 (corresponding to the rumen), C2 (corresponding to the reticulum), and C3 (corresponding to the fusion of the omasum and abomasum), although internally, the distal part of C3 is distinguished by the existence of mucus glands in which hydrochloric acid is produced for digestion, so that some describe a C4 compartment. Compartment C1 is itself subdivided into two "sacs," which in turn include rounded cul-de-sacs on their ventral side, called glandular sacs or water sacs (Fig. 4.6).

This anatomical configuration allows the animal to separate the food bolus in the stomachs into two phases, one solid and the other liquid. The consequence is that the ingested solid particles remain in the stomachs for longer. However, as in ruminants, digestion is done by the bacteria present. These bacteria will digest the cellulose of the plants and other components from which they will synthesize the proteins that will be absorbed in the intestine. The advantage of a longer stay of the solid particles in the C1 compartment will be a more complete digestion, which explains why large camelids are able to digest nutritionally poor forages in better proportions than cows (Jouany 2000).

In addition, the intestinal transit of nutrients from the C3 compartment to fecal excretion is slower than in other herbivores, which further increases the efficiency of digestion of poor forages.

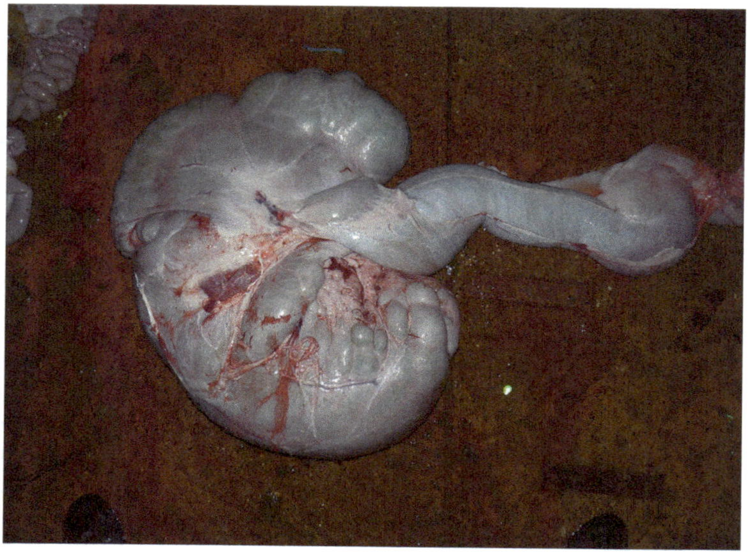

Fig. 4.6 Anatomy of the dromedary stomachs

However, two other characteristics of the species also contribute to the digestive efficiency of large camelids with significant consequences for feeding: the buffering capacity of stomach contents and the recycling of urea.

Buffer capacity is the ability of a milieu (in this case stomach contents) to maintain a stable level of acidity. If the buffering capacity is low, certain food rations can lead to an excess of acidity (acidosis) or alkalinity (alkalosis). However, these phenomena are rare in large camelids (except for racing camels fed with unbalanced rations very rich in concentrated feed). The buffering capacity is provided by the saliva that the camel produces in large quantities and by the phosphates and bicarbonates secreted by the glands located in the glandular sacs. The advantage of a more or less constant acidity (at least the pH is 6.5) is the maintenance of the activity of the bacteria in the stomachs. In addition, the separation of nutrients into two phases allows for a faster increase in the "turnover" of the liquid phase and finally, combined with the maintenance of bacterial activity, a faster elimination of fermentable sugars and products of fermentation that can otherwise cause digestive disorders.

The urea cycle is a classic phenomenon in herbivores. Indeed, plants contain proteins and nonprotein nitrogen. The digestion of proteins by bacteria gives amino acids, then urea, then ammonia. The nonprotein nitrogen will directly give ammonia. It is from ammonia that the bacteria will synthesize part of the proteins digested in the intestines. However, part of the ammonia passes into the bloodstream to be metabolized into urea by the liver. This urea passes back into the blood to be eliminated by the kidneys via urinary excretion. But some of the urea returns to the stomachs and saliva to become ammonia, which is used to synthesize microbial proteins. This is called the urea cycle. The recycling rate in camels is much higher than in other species: 94% in camels, compared with 75% in desert goats and 78% in desert sheep. It even rises to 97% in the dehydrated camel. The consequence is advantageous, since such recycling leads to the excretion of urine that is relatively low in urea (which perhaps explains its interest in traditional pharmacopoeia), and to greater digestive efficiency of low-protein fodder. On the other hand, this explains the risk of uremic poisoning in the case of a ration too rich in nonprotein nitrogen.

4.2.2 Nutrition of Large Camelids

Like almost all animals, large camelids need sugars, fats, proteins, minerals, and vitamins, in addition to water, to ensure their maintenance and production. They find all these elements in the forage available in desert or steppe ecosystems in varying proportions. Recommendations for nutrient requirements have been relatively unstudied in large camelids, and in general, for ration calculation (see Chap. 5 "Feed Management"), the references acquired in cattle are used while considering the specificities of digestion in this species mentioned above. In this chapter, we will focus on feeding behavior, intake, fat storage, and water requirements.

4.2.3 Feeding Behavior

Dromedaries and Bactrians are browsers rather than grazers, that is, animals that consume a wide variety of plants in their ecosystems, from the shortest grasses to the stems and leaves of trees accessible by their long necks (Fig. 4.7). In this respect, they are similar to goats and clearly distinguishable from cattle and sheep. Such behavior is linked to physiological and anatomical peculiarities: good digestion of cellulose, thanks to cellulolytic bacteria in the stomachs, strong secretion of hydrochloric acid in the C4 compartment, strong amylolytic activity allowing better digestion of starches, greater length of the intestine, and greater papillary surface.

When browsing, large camelids have a nongregarious and ambulatory behavior, which means a group of animals (a herd) never concentrates all together in the same place and that they move continuously to browse. These movements can be significant (more than 20 km per day), which avoids too much localized pressure on the resource. As browsing animal, its selectivity of consumed plants will therefore be higher. Numerous studies, particularly in Africa, have shown both the diversity of the species consumed and the variability of the plant strata concerned, thus avoiding the overexploitation of a given stratum. It has been estimated that a dromedary consumes between 15 and 22 different plants in a given environment, compared with only 12 to 15 for sheep (Rutagwenda et al. 1990). Overall, animals spend an average

a b

Fig. 4.7 Feeding behavior of grazing dromedaries, which consume everything from the shortest grass (**a**) to the stems and leaves of trees accessible by their long necks (**b**) of 8 hours on rangelands collecting forage resources for food. This behavior may have consequences in stall systems where feed is delivered to feed bunks, and rations are much more monotonous and distributed only twice a day.

4.2.4 Quantities Ingested

The quantity consumed by large camelids depends on their weight, age, forage quality, ration composition (presence or absence of concentrates), water content, and their status (growing animal, lactating or pregnant female, rutting, or resting male), which determines their needs. The figures given in the scientific literature are therefore very variable. On average, the quantities ingested by adults vary between 2 and 3 kg of dry matter (DM)/100 kg of body weight, i.e., for an adult weighing 500 kg, between 10 and 15 kg of DM. As bush grasses reach nearly 80% DM, this corresponds to about 12 to 18 kg of hay. The ingestion capacity is therefore generally lower than in cows, in the order of 18 to 22 kg DM for an animal of the same weight (500 kg).

4.2.5 Storage and Destocking of Fat Reserves

Large camelids are the only species able to store most of their fat reserves in one place, the hump (Fig. 4.8), which concentrates between 50 and 80% of the fat in the carcass (about 30–35% of total fat if intramuscular fat is included). Fat is stored in specialized cells called adipocytes. Fat accounts for over 90% of the hump's DM. The phenomenon of fat storage/destocking is essential for the survival of the animal. These reserves are stored during favorable seasons (rainy season) when grass

a b

Fig. 4.8 The fat reserves of the dromedary and the Bactrian are concentrated in the hump(s). The hump is composed of 70% fat is more abundant, or during periods of the physiological cycle that require less energy (end of lactation, beginning of gestation, period of sexual rest in males).

Conversely, these reserves, which constitute energy for the animal, are mobilized during unfavorable seasons, when food resources are of poor quality or when energy requirements increase (beginning of lactation, end of gestation, rutting period). The concentration of these reserves in the hump is a parameter of adaptation of large camelids to heat. Indeed, since the fat is concentrated in the hump, the subcutaneous fat on the rest of the body is not very abundant, which facilitates the dissipation of excess heat. However, there are two other internal storage sites, called perirenal fat located around the kidneys (about 10–15% of the fat) and mesenteric fat (5–10% of the fat), which covers the mesentery, the inner wall of the abdominal cavity. The advantage of the hump is that it gives an easy reading of the state of the fat stock in a given animal.

4.2.6 Water Requirements

The resistance of large camelids to thirst is proverbial. Indeed, the water requirements of the dromedary, notably, which have been studied for a long time, appear to be lower than in other domestic species living in desert regions. One of the parameters used to evaluate these needs is the water turnover, i.e., the need to renew the body's water. In dromedaries, this is approximately 80 ml/kg body weight per 24 hours. By comparison, it is 160 ml/kg/24 h in bulls, 120 in zebus, 100 in desert sheep, and even 95 in desert goats, which are resistant to dehydration. Water requirements depend of course on climatic conditions. They are usually higher in the hot season and with food rations richer in dry matter. It is estimated that water requirements are around 20–30 L/day (Box 4.1).

Although lactation is associated with increased requirements, several studies suggest that water restriction has a limited effect on milk production. However, it should be remembered that despite its ability to resist dehydration and to space out watering, the camel needs to drink like all living creatures.

Box 4.1 Dromedary Watering Patterns (Ethiopia, Sudan, Djibouti, Kenya, Saudi Arabia, Kazakhstan) (Fig. 4.9)

Fig. 4.9 Watering is practiced either by taking the animals to natural resources (river, lake) or arranged resources (troughs, containers at the exit of wells) or by bringing water to the herds (tanker)

Fig. 4.9 (continued)

References

Ayadi M, Aljumaah RS, Samara EM, Faye B, Caja G (2016) A proposal of linear assessment scheme for the udder of dairy camels (*Camelus dromedarius* L.). Trop Anim Health Prod 48(5): 927–933. https://doi.org/10.1007/s11250-016-1051-4

Aziz MA, Faye B, Al-Eknah M, Musaad A (2016) Modeling lactation curve of Saudi (camels using a linear and non-linear forms of the incomplete Gamma function). Small Rumin Res 137:40–46. https://doi.org/10.1016/j.smallrumres.2016.03.001

Jouany JP (2000) La digestion chez les camélidés : comparaison avec les ruminants. Prod Anim 13(3):165–176. https://hal.inrae.fr/hal-02694048

Musaad A, Faye B, Nikhela AA (2013) Lactation curves of dairy camels in an intensive system. Trop Anim Health Prod 45(4):1039–1046. https://doi.org/10.1007/s11250-012-0331-x

Nagy P, Juhasz J (2009) Intensification of milk production and machine milking of dromedary camels (Camelus Dromedarius). 2nd conference of the International Society on Camelids Research and Development, Djerba, Tunisia, 11–14 Mar 2009 (CD-Rom)

Rutagwenda T, Lechner-Doll M, Schwartz HJ, Schultka W, Von Engelhardt W (1990) Dietary preference and degradability of forage on a semiarid thornbush savannah by indigenous ruminants, camels and donkeys. Anim Feed Sci Technol 31(3–4):179–192. https://doi.org/10.1016/0377-8401(90)90123-P

Technical Management of a Camel Farm with a Focus on Dairy: Breeding and Young Animal Rearing, Milking, and Feeding

5

Abstract

Camel farm management involves taking care of the reproduction, milking, feeding, and health of all animals throughout their lives. Regarding reproduction management, the farmer has to control the mating, make pregnancy diagnosis, assist the delivery, take care of the baby camel at birth, and eventually apply biotechnology of reproduction. For milking either by hand or with machine, it is important to apply good milking practices, control the milk flow, achieve mastitis detection, and assume global milking hygiene. In addition, control of the growth performances of the young camels until weaning and beyond is necessary. Regarding the feeding system, the farmer must understand the requirements (protein, energy, minerals, and vitamins) according to the physiological status of the camel (lactating, pregnant, rutting, or nonrutting male, young in growth) and to take care of the availability of the resources and watering.

Keywords

Reproduction management · Milking practices · Feed requirements · Calculation of ration · Growth · Weaning practices

This part is devoted to the management of a camel farm with a particular focus on the management of a camel farm for milk production. As this is a delicate management, some basic rules for a good management of (i) reproduction, (ii) rearing of young animals, (iii) milking, (iv) feeding and finally, and (v) health will be established in the next chapter.

B. Faye et al., *Large Camel Farming*, https://doi.org/10.1007/978-94-024-2237-5_5

5.1 Reproductive Management

5.1.1 Management of Reproduction

Breeding of camels shortly after puberty or in later breeding cycles should be done on the first and second day of heat if visible, as it is when the proportion of fertile mating is highest. In case of silent heats, it is preferable to place the male within the herd of females to be mated. A reproducer can be used several times a day, and there is one breeder male for every 25–30 females in a herd, but it is considered that it is able to cover 40–50 females (or even 70) in total during a rutting season at a rate of 3 or 4 females per day. However, it is not recommended to keep the same male in a herd for more than 5 years to avoid too much inbreeding.

A female that is put to breeding too young weakens its fertility capital. Pregnancy will develop at the expense of its own growth, which will also compete with lactation. The time saved for the first birth may be lost by the time interval required by the young female to reconstitute her reserves before the second birth. In fact, more than an age criterion, which can vary between 2.5 and 5 years, the factor to be taken into consideration when deciding whether or not to breed is the weight of the young female: it must not be less than 65% of the adult weight.

In the traditional environment, the farmer waits until the camel is weaned before breeding the female again. However, as gestation causes a premature cessation of lactation (Nagy et al. 2015), mating too soon after calving is not advisable in dairy farms (see Chap. 3, Sect. 3.1). Under good maintenance conditions, females can be mated 4–5 months after calving if this occurs at the beginning of the male's breeding season, allowing a calving interval of around 16–18 months and obtaining two camels over 3 years. On-station trials have even been carried out to obtain early returns to heat by hormonal induction, allowing breeding to take place 1 month after calving and leading to intervals of 14 months.

Two types of breeding are described:

- Assisted mating consists of bringing the female to mate with the male having the desired genetic qualities in order to better manage the choice of breeding stock and to set up an appropriate selection of production or performance parameters. The advantage of this system, apart from the aspects concerning genetic improvement, is the possibility of preliminary control of the genital integrity of the reproducers (possibly with control of the follicles' maturity by ultrasonography), making it possible to optimize the success of the fertilization and to work under optimal conditions of hygiene. Mating takes place in a dedicated area where the breeder can assist the male during the intromission of the penis. Contrary to popular belief, mating can take place in the presence of a human, but it should be done in a calm environment, avoiding the presence of other adult males.
- Free mating is the one generally practiced in extensive environments. The breeding male is left permanently with the females in the herd during the breeding season. In more intensive production systems, it is brought into the pen of the females to be mated, either all day or only at night. This system requires little or

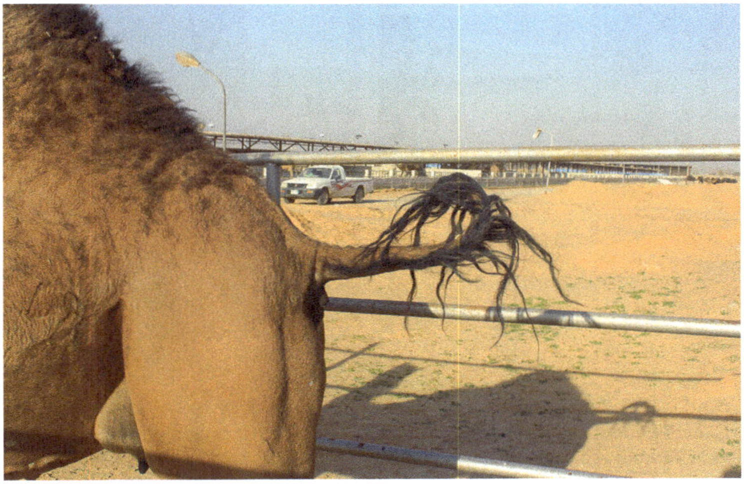

Fig. 5.1 Tail lift sign in mated females

no labor, but it leads to uncertainty about the precise date of mating. After a few weeks, mated females that show signs of pregnancy can be removed (see next chapter).

There are three diagnostic methods to confirm the pregnancy status of the female, from the simplest to the most sophisticated:

- The raising of the tail: this is the traditional method practiced by all breeders, as the fertilized female shows a fairly early behavior after fertilization, about 2 weeks after the start of gestation, which is manifested by a stiffening attitude with the head raised, and especially the lifting of the tail as soon as the male approaches (Fig. 5.1). The tail lift is not constant, but it occurs reliably in 95% of cases in calm camels. In contrast, there are many false-positive reactions in agitated females. This sign disappears rapidly after abortion or early embryonic death. In the Bactrian camel, this sign is less pronounced.
- The dosage of progesterone, the hormone of gestation. Its level in the blood, in case of a confirmed pregnancy, remains above 2 ng/ml, while in case of nonpregnancy after mating, it drops by 10 times, i.e., around 0.2 ng/ml. However, as the increase occurs immediately after ovulation and is maintained for about 8 days, even in the case of nonpregnancy, the assay for the diagnosis of successful fertilization is usually performed after 15 days postmating. The assay techniques used in laboratories for other species are perfectly valid for large camelids. It is also possible to measure progesterone in milk. There are few references in camels, probably because breeding often takes place after drying off, but levels vary around 5 ng/ml in pregnant females and above compared to less than 1 ng/ml of milk in nonpregnant females.

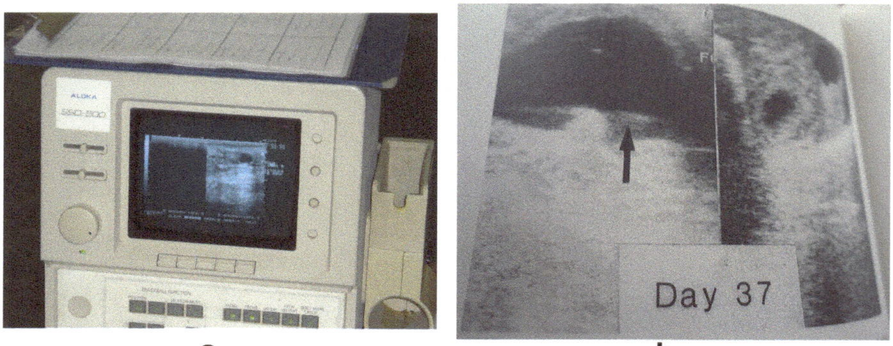

a b

Fig. 5.2 (**a**) Portable ultrasonograph used for pregnancy diagnosis or any gynecological examination of large herbivores such as camels. (**b**) diagnosis of pregnancy by ultrasonography at 37^{e} days of gestation in a camel. The arrow indicates the allantoic membrane. (Source: Tibary and Anouassi 1997)

a b

Fig. 5.3 Rectal palpation. (**a**) In a restraint cage with an ultrasound machine; (**b**) on an animal in a prone position

– Rectal ultrasonography: this has been a useful replacement for manual rectal palpation, which requires more practice. There are now many easy-to-use portable devices available (Fig. 5.2). It can be performed with high reliability as early as 18^{e} days of gestation.

In the absence of an ultrasound machine, it takes some practice to make a diagnosis by transrectal palpation alone. If a suitable restraint cage is not available (see Sect. 6.1), the animal must be restrained. The animal does not accept this handling well, which may require the use of a tranquilizer (Fig. 5.3).

In all cases, the signs that are noticeable during gestation are as follows:

- 1st month: presence of a corpus luteum on one of the ovaries
- 2nd month: uterus still in the pelvis, left uterine horn evenly enlarged, corpus luteum either on the right or on the left ovary
- 3rd month: left uterine horn clearly larger than the right horn, ovary in the abdominal position
- 4th month: cervix at the edge of the pelvis, uterus still fully palpable at the edge of the pelvis
- 5th month: uterus in the abdominal position, fetus sometimes palpable
- 6th month: dorsal wall of the uterus slightly below the floor of the pelvis, ovary of the right horn still palpable
- 7th month: uterus under the floor of the pelvis but still palpable, head and legs of the fetus identifiable
- 8th month: fetal head, neck, and front legs palpable
- 9th month: fetus becomes detectable and can be swung to the right side
- 10th month: obvious movements, beginning of mammary development, the ovary of the right horn becomes difficult to detect
- 11th month: onset of udder hypertrophy and vulvar loosening
- 12th month: abdominal perception of gestation evident, caudal part of the uterus occupies the front of the pelvis, relaxation of sacroiliac ligament
- 13th month: vulva swollen, udder developed, fetus may swing toward both abdominal walls

5.1.2 Assistance to the Parturition

In most cases, the camel gives birth easily, and dystocia (difficulty in giving birth) is rare (less than 5% of births). It is common in extensive systems to find the camel close to the mother without the need for human intervention. In general, dystocia is considered to occur when the preparatory phase lasts more than 6 hours and the expulsion phase more than 2 hours. The female then shows signs of distress and keeps changing position, alternately standing, or lying down. Two types of dystocia are described: those of fetal origin and those of maternal origin:

- Dystocia of fetal origin: in the first case; the position of the fetus should be examined, i.e., its orientation in the matrix, whether it is presented by the head and the front limbs ("normal" position), or with the front limbs folded backwards, or whether it is presented by the rear ("sit" position), with the hind limbs folded under it or not, or with the neck folded back, etc. Difficulties in calving may also be associated with monstrous anomalies, which are very rarely described, except in the heavily polluted areas of Central Asia on Bactrian camels.
- Dystocia of maternal origin: the most frequent cases are due to uterine torsion that can generally only be resolved by caesarean section. Dystocia can also occur in primiparous females that are put to reproduction too early and whose pelvis is insufficiently developed to let the fetus pass, or conversely in old females affected by uterine atony, but these cases are rarer.

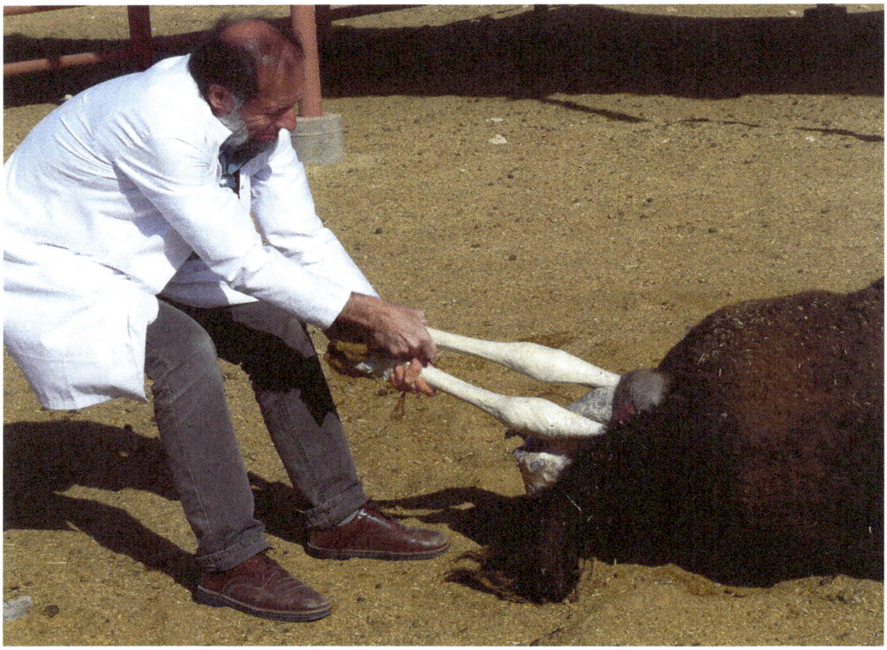

Fig. 5.4 Parturition assistance in a primiparous camel. (Photo G. Konuspayeva)

In all cases, it is best to call a veterinarian who will attempt to return the fetus to a more favorable position (by pushing or rotating) in fetal dystocia (Fig. 5.4), or to perform a Caesarean section in maternal dystocia.

Caesarean is performed when the fetus/pelvis ratio does not allow the young to be expelled by the natural route (too large calf or too narrow pelvic canal), in case of uterine torsion, malformation of the calf, fetal emphysema, or poor position of the fetus in the uterus impossible to reduce by external manipulation. Caesarean is performed on a female in the sitting position, by incision on the left side after an injection of a tranquilizer and local or regional anesthesia. Different techniques are described in the literature.

5.1.3 Care of the Newborn Camel

The birth of a camel represents a profound change for any species that should be evaluated to limit mortality in young age, which represents the most important economic loss in camel breeding. From birth, it is necessary to assess the respiratory functions of the camel and its vitality as well as its ability to stand up quickly and suckle its mother very quickly. The chances of survival of a camel lacking tone and weighing less than 25 kg (showing its immaturity) are much lower. Respiratory distress is easily visible (open-mouth breathing, rales). The suckling reflex normally

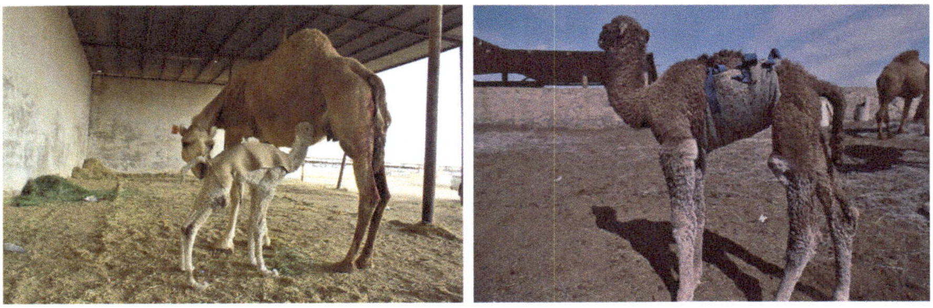

Fig. 5.5 Camel's protective blanket against cold in Saudi Arabia (**a**) and Kazakhstan (**b**)

appears within the first half hour. The epithelion is translucent in normal parturition but becomes yellowish in prolonged dystocia. From the first attempts of the camel to stand up, it is possible to observe possible anomalies such as deformities of the limbs causing abnormal postures, which usually resolve within a few weeks. Their cause is often associated with inbreeding problems.

The main mortality factor is cold. Indeed, the breeding season for males corresponds to winter, and gestation lasts about 13 months, so most births occur during the coldest months of the year. It is therefore advisable to protect the young animal as much as possible in a dry place if the birth takes place in nondesert ecosystems. In some countries or areas where winters are particularly harsh, the use of blankets adapted to the anatomy of the camel is recommended (Fig. 5.5).

Three types of precautions should be taken to protect the newborn from potential infection or mineral deficiency:

- Vaccination and treatment: to avoid the risk of enterotoxemia, vaccination of the mother against *Clostridium perfringens* may be recommended, as well as vaccination against tetanus or specific viral diseases (e.g., IBR-BVD). Injection of the vaccine at the end of gestation (last month) allows the protection of the newborn, thanks to the transfer of antibodies to the fetus. In some regions of the world (particularly in the Gulf States), selenium deficiency is one of the major causes of mortality in young animals during the first month of life. The injection of a solution of Selpherol®, e.g., or the administration of boluses containing selenium makes it possible to ensure a sufficient rate in the young camel during the first weeks of its life, contributing to the prevention of the risk of cardiac degeneration.
- Disinfection of the umbilical cord: as soon as the baby is born, it is recommended to cut and disinfect the umbilical cord with an iodine solution, as it can be an important entry point for environmental pathogens. If the birth takes place in a closed building, it is recommended that it is clean, especially free of animal excrement.
- Colostrum intake: Colostrum is the first milk that allows a passive transfer of immunity via immunoglobulins (IgG), which do not cross the placental barrier

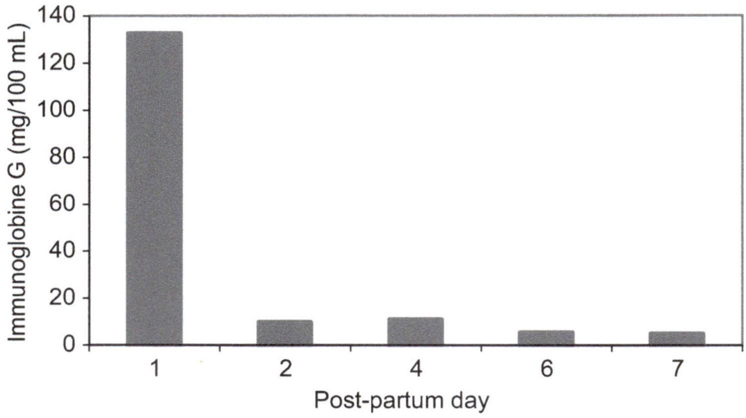

Fig. 5.6 Immunoglobulin G concentration (in mg/100 ml) in camel colostrum in the first 7 days of life. (Source: Konuspayeva et al. 2007)

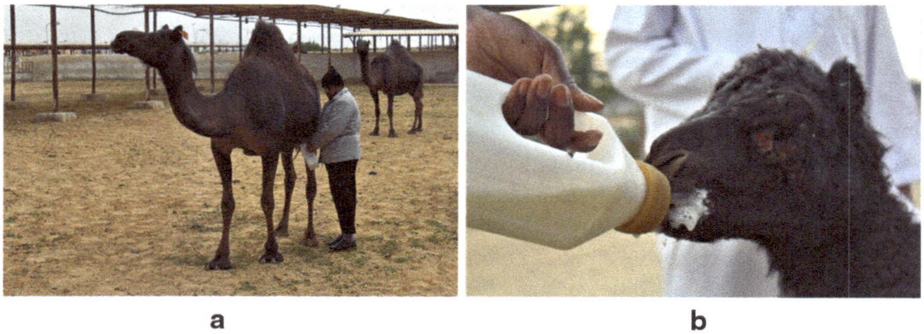

Fig. 5.7 (**a**) Milking a camel that refuses her calf; (**b**) distribution of colostrum to a rejected camel

during in utero life. In addition, IgG levels decrease rapidly after the first milk is ejected (Fig. 5.6). It is therefore recommended that the calf should be able to suckle its mother, if possible, within the first hour after birth to take advantage of the high IgG levels in the colostrum and to take advantage of the maximum intestinal absorption capacity provided during the first 24 hours. On average, it is considered that the calf can suckle about 200 ml of colostrum every 2 hours during the first 18 hours. If the camel is too weak, the camel can be milked, and the colostrum can be given by bottle.

It is therefore essential for the breeder to monitor the first few hours after birth to ensure that the colostrum is taken and that the newborn has sufficient vitality to stand and access the maternal teats. It is also possible to provide colostrum from another good-producing camel, frozen for distribution to weaker camels or from a camel that is refusing her calf (Fig. 5.7).

5.1.4 Adoption Techniques

One of the reasons for the mortality of the young is the inability of the mother to suckle her calf, either because of the refusal of the camel, or because of lack of milk, or for any other reasons (painful suckling or death of the mother). In order to save the camel, pastoralists have developed adoption practices that are more or less effective and aim to have the camel accepted by its own mother or by another lactating camel that has lost its young.

The most common technique is to present the mother with a foreign camel calf wrapped in the placenta of her stillborn calf (Fig. 5.8).

Other adoption techniques based on odor recognition include blindfolding the mother and placing the skin of the stillborn baby camel on the substitute calf, which is presented to the mother when the blindfold is removed.

There are also more coercive traditional practices. Among the Tuaregs, for example, a real psychodrama is carried out: the substitute camel is tied to the adopted female, whose eyes have been blindfolded beforehand, and then the breeders, aided by dogs, simulate an aggression with shouts, barking, sometimes blows, and projectiles; the distraught baby camel tries to take refuge with the female, who, in turn, out of an instinct for self-preservation, tries to protect the youngster she feels close to her. After several tens of minutes of such a disturbance, the female's blindfold is removed while leaving the substitute camel attached to her. Conversely, in Central Asia, there is a less violent technique for cases of self-adoption (when the mother refuses her baby), which consists of provoking an emotion in the mother by playing soft music at the moment of presenting her baby.

Some camel drivers also perform a mock birth. This particularly aggressive technique consists of introducing a bag filled with camel dung into the female's anus, then sewing her anus shut to prevent fecal excretion for 3–4 hours. In addition to this technique, the nostrils are blocked by placing a plastic bag inside them before tying them up. After 3–4 hours, the anus is cut open, thus giving the mother the illusion of giving birth. At the same time, the ties obstructing the female's nostrils are

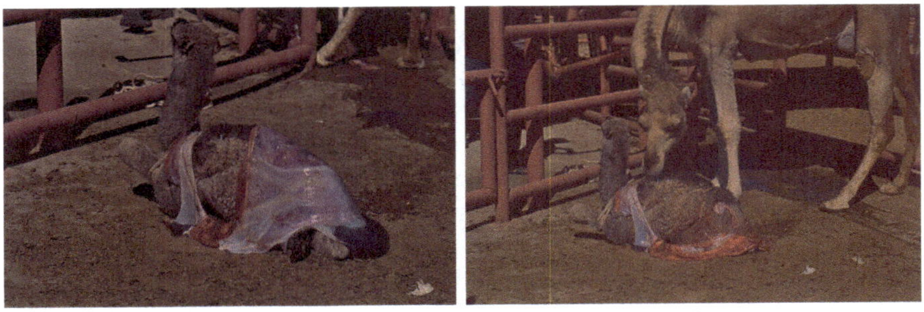

a b

Fig. 5.8 (**a**) Baby camel covered with the placenta of another stillborn baby camel for adoption; (**b**) acceptance of a foreign calf by a female camel that has lost her calf in parturition undone before she is presented with her calf or a substitute camel, which she smells as if she has just given birth (Fig. 5.9). These techniques are generally efficient.

5.1.5 Management of Breeding Males

The number of breeders are few in a herd. However, it is important to manage them correctly so that they can perform their function. Several parameters must be taken into consideration for a good management:

- The age of the reproducers: the ideal age is between 6 and 12 years old; too young, the weakening of the rutting period does not allow them to ensure their reproductive function; too old, their fertilizing power may be limited, and their preferences may lead them to abandon certain females.
- Conformation: the selection of the sire is the first factor of genetic improvement of the herd. In the absence of dairy indexes, breeders can differentiate between those that pass on their conformation to their offspring. One of the empirical criteria used for sires in dairy farms is the size of the penis sheath, which is correlated with the mammary development of female offspring.
- Genetic origin: a sire from outside the herd is a guarantee of a certain genetic variability, beneficial to improvement. A male breeder from the same herd as the females to be mated, especially if kept for too long (more than 5–6 years), risks mating related females, leading to inbreeding which can cause sterility or malformations. It is therefore advisable to replace the breeding stock regularly.
- The number of breeding animals: it must be in accordance with the size of the herd. In general, we recommend one male for every 30–35 females. In large herds, the cohabitation of several males can cause problems of aggressiveness during the rut, which is why the breeders are isolated from each other. However, it is possible to keep a young male next to a dominant adult: this one avoids direct confrontation but can cover the females left behind by the first one.
- Maintaining libido: in a sedentary context and although isolated, the male develops a libido that is better expressed if it remains in the visual presence of females, especially during the rut. It is therefore preferable to place it in an enclosure that allows it to be visually close to the females. In an extensive environment, the male is left with the herd of camels to be mated, but it should be left alone as a breeder in his harem.
- The diet: it is to be taken care of especially during the sexual resting season. A good food supplementation allows the reproducer to reconstitute his reserves before facing the next rutting period during which it is able to lose 20–30% of his weight.
- Castration of undesirable breeding stock: castration is reserved either for aggressive males, or for males intended for activities other than reproduction (work, riding), or to avoid aggression between males in a herd or to avoid unwanted mating. In general, castration is performed at around 2–3 years of age, although earlier (before 6 months) or later (up to 5–6 years) castrations are possible. There

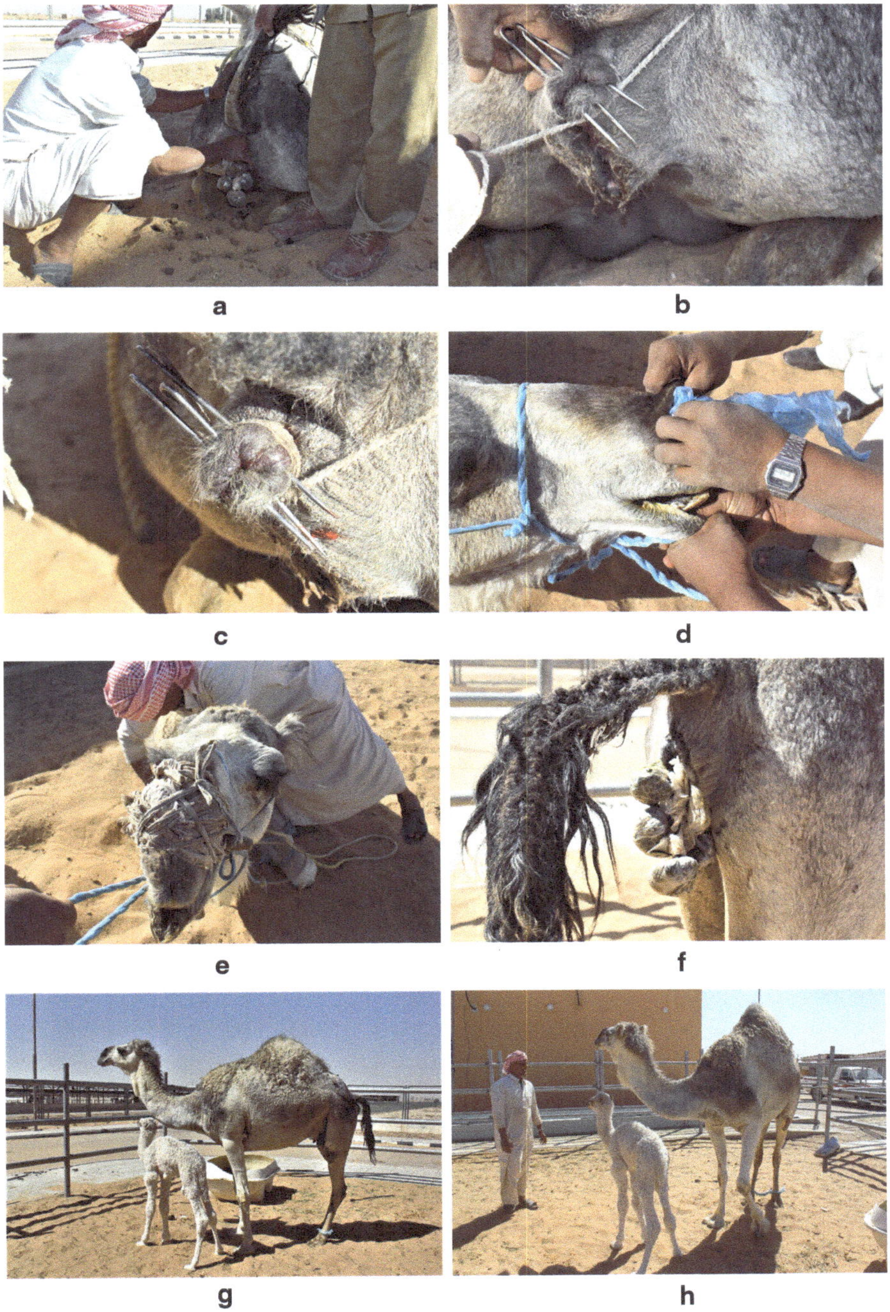

Fig. 5.9 Traditional adoption technique simulating childbirth. (**a**) Introduction of bags filled with dung into the anus; (**b**, **c**) ligation of the anus; (**d**) introduction of plastic into the nostrils; (**e**) ligation of the nostrils; (**f**) release of the anus; (**g**) presentation of the calf after release of the nostrils; (**h**) recognition of the calf by the mother

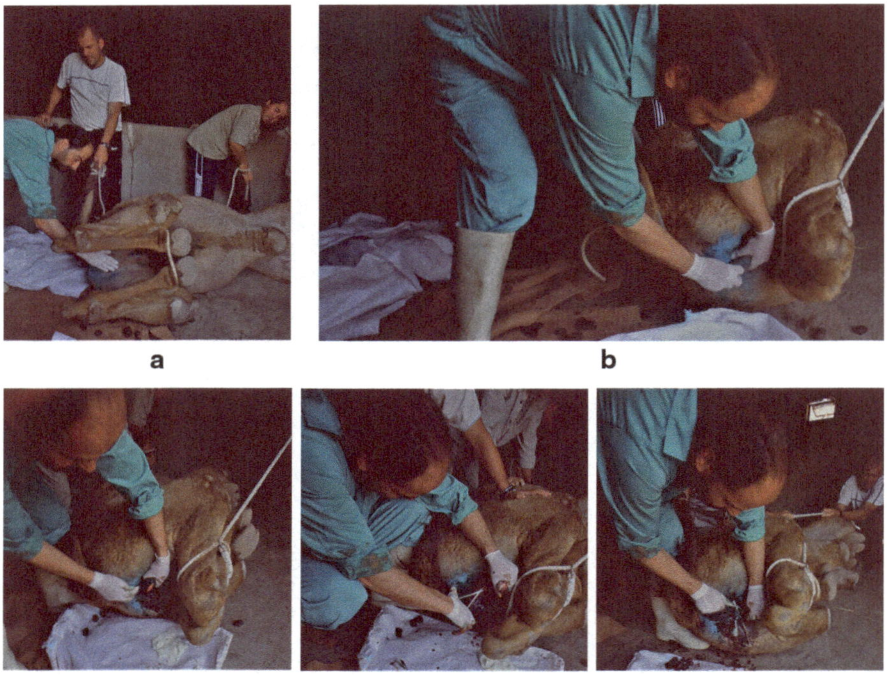

a b c d e

Fig. 5.10 Castration of a camel. (**a**) Restraint and disinfection of the area; (**b**) incision on the median raphe of the testis; (**c**) extraction of the testis and removal; (**d**) clamping of the testicular cord; (**e**) skin suture of the scrotum

are several techniques for castration, but the most common includes the classic steps (Fig. 5.10): (i) local anesthesia after sedation; (ii) restraint of the animal; (iii) disinfection of the bursa; (iv) incision on the median raphe of 6–8 cm; (v) exteriorization of the testis, covered or not; (vi) transfixing ligation with absorbable thread; (vii) repeat with the second testicle; (viii) possibly cutaneous suture of the scrotum with an absorbable thread; and (ix) disinfection of the wound and application of an insecticide repellent such as Tigal around the wound, to be repeated every day during the week following the operation.

5.1.6 Reproductive Biotechnology

Sperm collection and freezing, artificial insemination, and embryo transfer are possible in large camelids and are practiced in several places in the world. The interest in using these techniques is multiple: reproductive hygiene, prevention of sexually transmitted diseases, grouping of breeding and births, and especially genetic improvement. However, none of these reproductive biotechnologies could be developed on a large scale due to both technical constraints and organizational limitations.

5.1.6.1 Artificial Insemination (AI)

AI requires the collection of semen from the male. For such collection, the electro-ejaculation method, considered too coercive for the male and incompatible with animal welfare, has been progressively abandoned in favor of the use of the artificial vagina, of which several models exist. For the collection itself, two methods are advocated: (i) placing the artificial vagina when the male begins to straddle the female in heat that is presented to it; (ii) using a dummy. The first method requires the presence of an experienced technician as the duration of copulation and the position of the mating make the posture of the technician carrying the artificial vagina uncomfortable and potentially dangerous (Fig. 5.11).

The dummy technique was developed in Egypt. However, this technique requires good training of the male to accept the dummy and proper setup as the dummy is placed on a platform under which the collection operator with an artificial vagina is located (Fig. 5.12).

However, the main constraint to the development of AI in large camelids is its consistency. Indeed, the semen of the male is very viscous and therefore difficult to manage. Moreover, the spermatozoa of large camelids do not tolerate freezing well, and it is therefore difficult to use frozen straws as it is done for other species. Much research is being done on diluting semen to reduce its viscosity or making it freezable while maintaining sperm motility and survival. In most cases, insemination is therefore performed with fresh semen. These constraints explain the relatively low performance of AI to achieve pregnancy (around 30–35% in dromedary females, slightly higher in Bactrian females). As a result, the technique has hardly left the research stations and a few large farms in the Gulf countries.

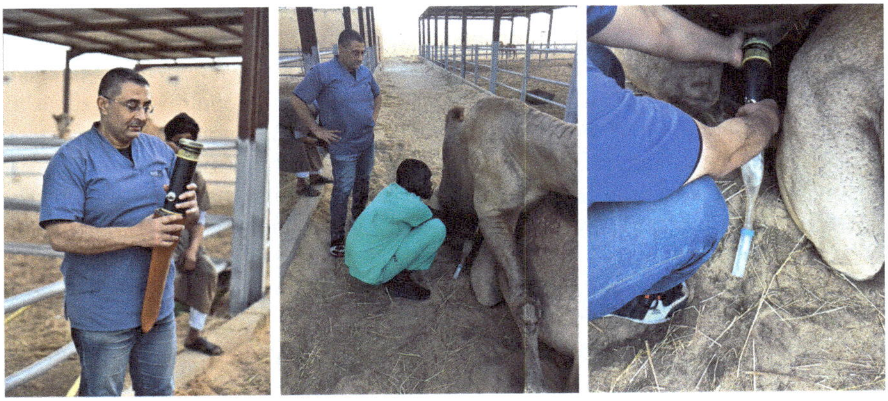

a b c

Fig. 5.11 Technique of semen collection with artificial vagina during natural mating

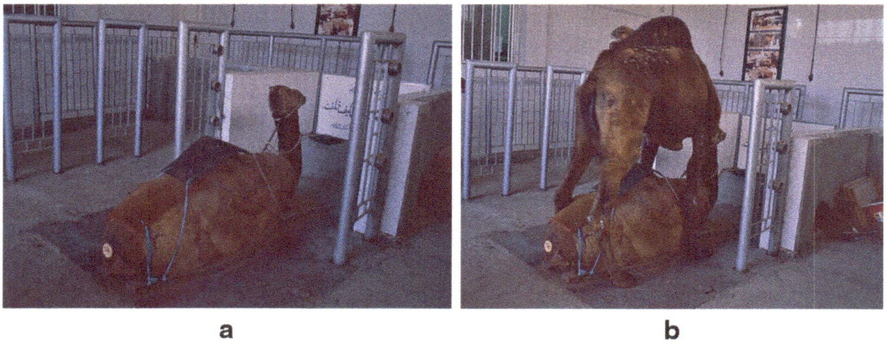

Fig. 5.12 Mannequin technique for collecting male camel semen in Egypt. The operator stands below the dummy in an appropriate room

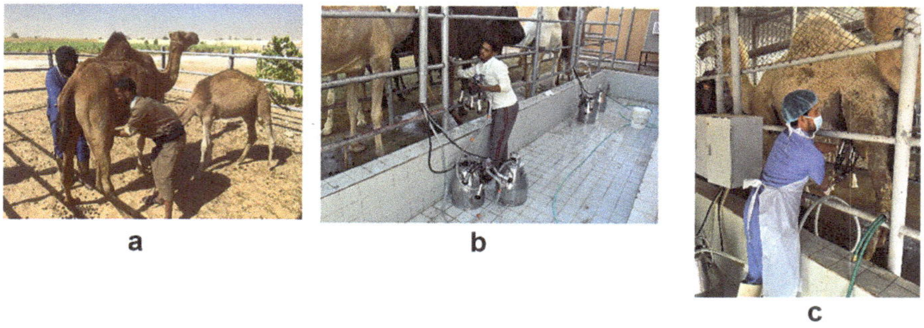

Fig. 5.13 Three levels of hygiene in milking practices. (**a**) Low level; (**b**) Medium level; (**c**) High level

5.1.6.2 Embryo Transfer (ET)

The ET technique appears to be more operational than AI, although it faces the same constraints as semen collection. The cost and technicality of ET methods relegate their use to large dairy farms and racing stables in the Gulf countries, as well as to research stations around the world. Although the protocols for the various steps (superovulation hormone treatment in donor females, natural mating, embryo collection, hormone synchronization in recipient females, embryo transfer, pregnancy monitoring) are now fairly well codified, it remains a technique that is not available to everyone (Fig. 5.13).

However, the success rate in terms of fertilization is higher than that of AI and allows even faster genetic progress, as a single donor can provide an average of seven embryos. Moreover, because embryos are easier to freeze than semen, the marketing of frozen embryos is easier than that of semen straws. Currently, only the United Arab Emirates markets frozen camel embryos with genetic improvement potential for milk production or racing.

5.2 Management of Milking

5.2.1 Presence of the Camel Calf

According to the breeders, the presence of the camel is absolutely necessary to ensure the descent of the milk. From this point of view, several practices are used. The farmer lets the youngster suck two of the teats on one side while milking is started on the other two teats; or the calf starts the milk production by the mother, then the farmer restrains the calf nearby (so that the youngster does not disturb the milking), takes the desired amount of milk, and then lets the young suckling the remaining milk (Fig. 5.14) The suckling activity of the camel is essential to ensure a higher milk production. For example, in a traditional setting, it has been shown that the daily amount of milk produced is 65% higher in a female camel with a calf surviving up to weaning than in a female camel that has lost her calf before weaning. Over the entire lactation, production can be 2.9 times higher if the calf survives.

The suckling activity of the calf provokes neurohormonal reflex leading to the secretion of oxytocin, the hormone provoking the opening of teat sphincter. However, the implementation of more intensive systems shows that prior suckling of the calf is not necessarily essential (Fig. 5.15).

The separation of the calf after weaning (see Sect. 5.3.2) requires that their role in oxytocin stimulation is supplemented. There are two possible techniques: (i) leaving the calves in the presence of the female during milking without access to the maternal teats, e.g., by placing a gate (Fig. 5.16) between the milking corridor and the calves' pen; (ii) getting milking females to be accustomed to the absence of the calves by manual stimulation (teat massage). Trials showed that optimal production was achieved after a massage of 2 minutes, which is much longer than for the cow (Ayadi et al. 2016).

In some countries, 1–2 ml of oxytocin is injected intramuscularly (IM) just before milking to replace the heifers and ensure milk let-down. However, this practice is not recommended if it is repeated regularly. In addition, the consumption of milk

a b

Fig. 5.14 (**a**) Traditional milking in the presence of the camel in Turkmenistan. (**b**) Milk sharing between man and calf during traditional milking in India

Fig. 5.15 Milking camels without calves at Camelicious Farm in Dubai

containing traces of synthetic oxytocin may have adverse effects on humans. The use of oxytocin should only be recommended in case of loss of the calf to stimulate milking after the death of the calf or when the female refuses to give milk during the first milkings.

5.2.2 Pace of Milking

In the traditional environment, the camel is milked twice a day, in the morning before leaving for the pasture and in the evening when returning to the camp, but, if necessary, the shepherd can collect milk every 2–3 hours. Increasing the frequency of milking has an impact on production. Increasing the frequency of milking from two to four times a day would increase production by 12%. However, milking can also be very irregular, with camels sometimes going 24 hours or more without being milked by the shepherd. Such irregularity in milking frequency can have negative consequences both on production and on the risk of mammary infections.

In an intensive environment, multiple daily milking sessions are a labor problem anyway, and the expected increase in production does not compensate for the additional staff costs and may exhaust the lactating female. The milking interval is also an important issue. Milk quantity decreases when milking intervals exceed 12 hours, and the optimum production would be reached after 8 hours between milking. It is therefore recommended to milk twice a day with an optimum milking interval of 8–12 hours maximum during the day, e.g., 7 am and 4 pm.

Fig. 5.16 Setting up a separation grid between the milking corridor and weaned camels

5.2.3 Types of Milking

As with other dairy species, a distinction is made between manual and mechanical milking:

– *Manual milking*: the most widespread in traditional environments, but also in semi-intensive systems, particularly in periurban livestock. Manual milking is more labour intensive. The milker (male and/or female, depending on the culture) stands on the side of the animal, with the calf on the opposite side (see Sect. 5.2.1). Milking can be done with one hand to hold the milking container with the other hand, or if the container is held by a second operator or attached to the neck of the milker (Fig. 5.17a), milking is done with both hands and is faster. In the case of high-yielding camels, two milkers placed on each side of the animal can achieve good results (Fig. 5.17b). In all cases, the experience of the milker is an important factor to ensure that milking is done in the best possible way (skill, milking speed, good udder emptying). Milking containers are varying widely in traditional farms, and some materials (calabashes, woven baskets, plastic containers) are not compatible with the recommended hygiene. However, metal containers, which are easier to clean, are more and more frequently used,

Fig. 5.17 Manual milking. (**a**) Traditional milking in Turkmenistan with a bucket attached to the neck; (**b**) manual milking in Kazakhstan of a good producing camel by two milkers simultaneously

- *Mechanical milking*: the use of the milking machine in camels dates back to the 1960s in the Soviet Union, but its development in other parts of the world is quite recent. Several types of equipment have been tested, namely, the individual independent milking pot, the individual milking pot connected to a central vacuum and pulsation system, and a centralized collection system (Fig. 5.18).

Regardless of the milking system and equipment, the most important factors to consider are vacuum and pulsation parameters. Several studies show that a vacuum of about 50 KPa and 60 pulsations/min are optimal for milk ejection, which are significantly higher than in cows (Ayadi et al. 2018). However, there is little data on the long-term effects of such a high vacuum. In farms where camels are selected to give milk easily, a vacuum of 45 KPa might be sufficient. Another point is related to teat shape (see Sect. 4.1.1). In ecotypes with poor mammary development and small teats, conical goat liners may be more suitable. In any case, care should be taken to ensure that the liners fit the shape of the teat to avoid long-term damage to the udder (Fig. 5.19).

5.2.4 Milk Ejection and Milking Quality

The use of a milk flow recording system (see Sect. 4.1.2) allows access to many parameters that can be used to interpret milking difficulties and the behavior of camels during milk ejection, and therefore the quality of milking. These parameters vary from one type of equipment to another, but globally, the following information can be accessed: Electrical conductivity (which allows detection of udder infections); total milking time from the placement of the milking machine liners to their removal (in minutes); total amount of milk produced during milking; duration of milk ejection (slightly shorter than the duration of the whole milking); duration of the different phases (up, plateau, down); average flow rate calculated during the milking time (in kg/min); maximum flow rate (in kg/min); presence or not of a bimodal flow; duration of overmilking (period between the end of milk ejection and

Fig. 5.18 Types of mechanical milking system. (**a**) Individual milking pot; (**b**) individual milking pot with centralized vacuum and pulsation system; (**c**) centralized collection system. (Photos B. Faye and A. Musaad)

the removal of the claw of the milking machine); duration of draining (milk obtained by increasing the manual pressure on the claw at the end of milking); and average flow rate of the drained milk and the quantity of milk produced during draining.

In any case, the interpretation of the milk flow curves can only be done if the milking of the camel takes place in optimal conditions of well-being. Stress (too much noise, sudden movements of the milker, number of people not usually present, untimely handling of the animals) should be avoided before and during milking. Feeding concentrates during milking can be a positive factor but is not mandatory.

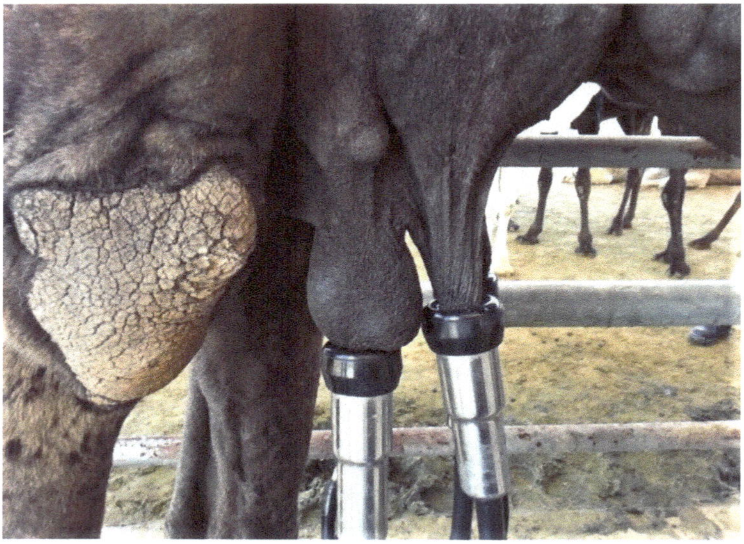

Fig. 5.19 Poor fit of liners to teat shape in a dairy camel

In general, a curve characterized by a rapid rise in milk flow to the plateau phase and a gradual downward phase is characteristic of optimal preparation as shown in Fig. 7.3. A break in the rising phase or plateau phase may indicate an incorrectly positioned claw of the milking machine. A bimodal curve indicates a low readiness to milk (insufficient stimulation) with a gap between cisternal and alveolar milk. A curve characterized by a dome-shaped plateau phase suggests that the preparation for milking was not sufficient. A low average flow rate is due to "difficult to milk" camels. A long decreasing is often associated with overmilking of individual quarters, expressed by larger or smaller steps in the curve. Too small teats can lead to air ingress, which can lead to liner or even whole claw drop, resulting in sudden breaks in the milk ejection curve. The fall of the claw can also be caused by nervous camels, trampling in the milking corridor, perhaps in connection with pain in the udder. In a study carried out in Tunisia on a large number of milk ejection curves, three main types of curves were observed (Fig. 5.20).

The analysis of the milk flow with a recording device can therefore help identify the camels that are easy to milk and, based on this observation, to establish a selection plan. In any case, milking under good conditions (well-trained camels, nonstressful milkers, good hygiene conditions) guarantees better productivity.

5.2.5 Training Camels in Mechanical Milking

The first step in achieving good quality milking is to familiarize the camels with the milking machine. There are different parlor configurations, with or without a milking

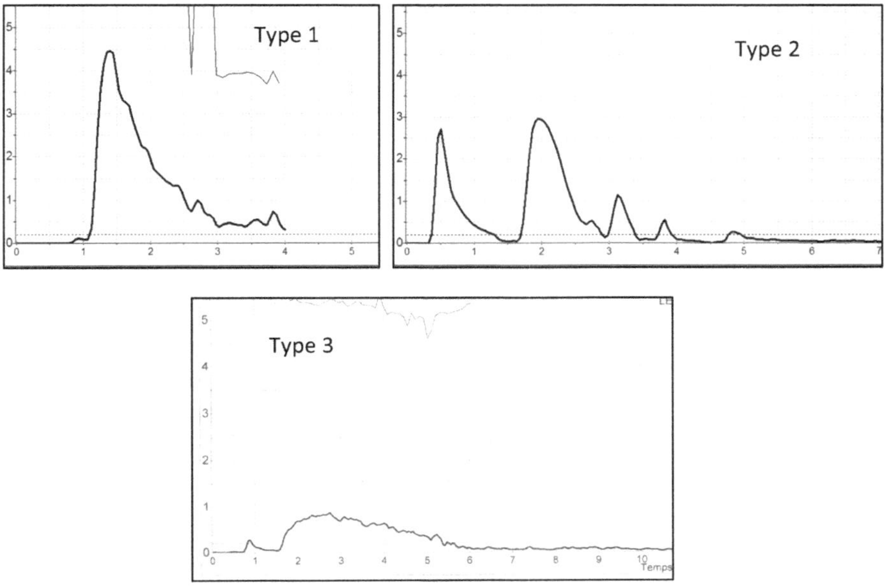

Fig. 5.20 Types of milk ejection curves. Type 1: a camel with a teat sphincter of low resistance allowing a rapid descent of milk; type 2: a camel with a resistant sphincter or with insufficient preparation or maladjustment of the teat liner; type 3: a stressed camel, poorly adapted to mechanical milking. (Source: Atigui et al. 2014)

corridor (especially in the case of individual milkers), linear, or herringbone corridor. The first step is to familiarize the camels with the milking environment, even before they are available for milking, i.e., during the first weeks when all their milk is still destined for their calf. The first difficulty lies in getting them to enter the milking corridor (the same difficulty arises in getting them to enter a holding pen). Coercive methods with the use of sticks are to be avoided to limit much stress and specially to avoid that the animal associates milking with pain. The camel calf can be used to induce it to enter the corridor and be placed among the females already trained (Fig. 5.21). Repeat this procedure for several consecutive days if the camel refuses to enter the corridor the first time. As soon as the camel accepts, offer a "reward" if possible (a little concentrate or a delicacy) and reassure her.

If there is a problem with fitting the liners (the camel may hit the claw with its hind legs, may bite the milker), it is safer to restrain the hind legs and have an assistant prevent it from biting (Fig. 5.22). This restraint is removed as soon as the camel is accustomed to the presence of the milking machine.

The duration of training varies from one camel to another and depends on it temperament, but it takes about 8–10 days for the right reflexes to be established and for the clawing to be accepted. After this time, the sound of the milking machine alone is enough for the camels to move towards the milking area. It is important that milking sessions are regular, at a fixed time and that the order in which the camels

Fig. 5.21 Training camels to enter the milking corridor

enter the corridor is respected (often the same camel enters first). Changes in acquired habits are often not accepted. For example, if there are two milking lanes on each side of the parlor, there is no point in trying to change the camels' lanes. The more acceptance there is for the actions and practices, the easier it will be to apply the hygiene rules.

5.2.6 Hygiene of Milking

In the traditional environment, hygiene rules are often reduced to the strict minimum: washing the milker's hands and cleaning the udder with water. In intensive farms and especially with the use of the milking machine, stricter rules should be imposed. Milk is a highly perishable biological liquid, which can be contaminated by pathogenic bacteria and other physical and chemical products. Mammary infections and milk contamination can be largely avoided by good milking practices. Good milking practices are based on 5 Ms: hygiene of the *milieu* (environmental hygiene), *material* hygiene (equipment), *manpower* hygiene (labour), *milking method hygiene*, and *primary matter* hygiene:

– *Environmental hygiene*: milking should be done in a clean area before the animals arrive and cleaned up immediately after milking. This includes removing accumulated feces from the milking area between milkings, not spilling mastitis

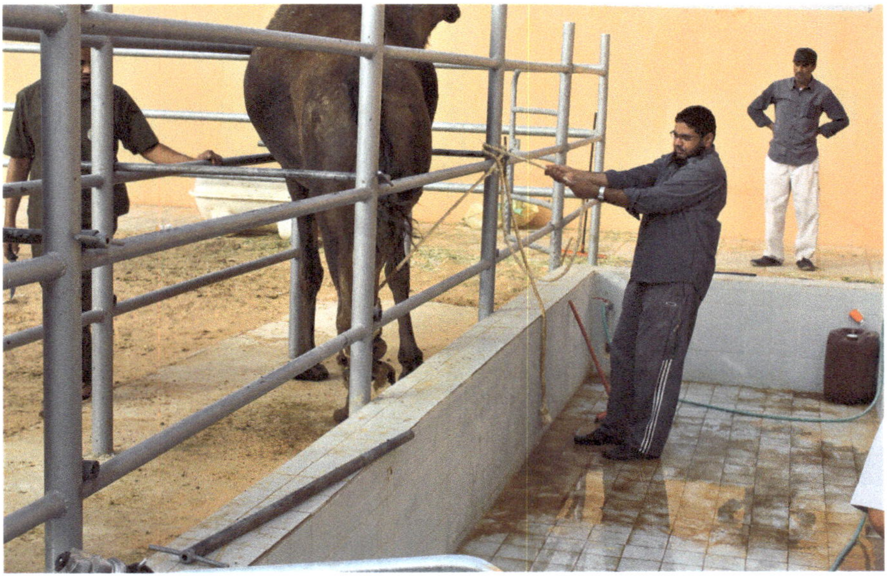

Fig. 5.22 Hindering a camel in the milking corridor before applying the liners

milk on the floor if it occurs, and avoiding the proximity of other animals during milking (dogs, small ruminants, birds).

- *Hygiene of the equipment*: the milking containers must be clean, cleaned with soapy water between milkings, and allowed to dry. If the milking machine has a large number of tubes, cleaning is based on a strict protocol that includes an acid and basic wash followed by a rinse (Fig. 5.23a). The milk should be stored in clean and disinfected containers between each use, if possible, in stainless-steel elements which are easier to clean. Storage and transport containers with small openings should be avoided, as they are more difficult to clean. A filter should be used before pouring the milk into cans (if there is no centralized collection system), and care should be taken to ensure that it is clean (Fig. 5.23b). If milking is done in the presence of the camel, the attachment ropes should also be cleaned regularly.
- *Labour hygiene*: the milker should wash his hands with soapy water before milking or wear gloves. The milker should wear clean clothes, possibly with a specific garment for milking. In milking parlors, it is even advisable to wear a mask, cap, and washable apron. It is especially not advisable to milk if you are ill or have sores on your hands,
- *Hygiene of methods*: before ejecting the milk, the milker should clean the teats with clean water or, better still, with soapy water, to remove bacteria from the skin and use a clean individual cloth to wash and dry the teats. The wipes can be disposable or washable (Fig. 5.24). In the latter case, the used wipes are placed in a dedicated box, and for each new camel milked, a clean wipe is taken from a

Fig. 5.23 Equipment hygiene. (**a**) Cleaning milking equipment after the animals have passed through; (**b**) use of stainless-steel containers and clean filters when pouring milk into storage or transport containers

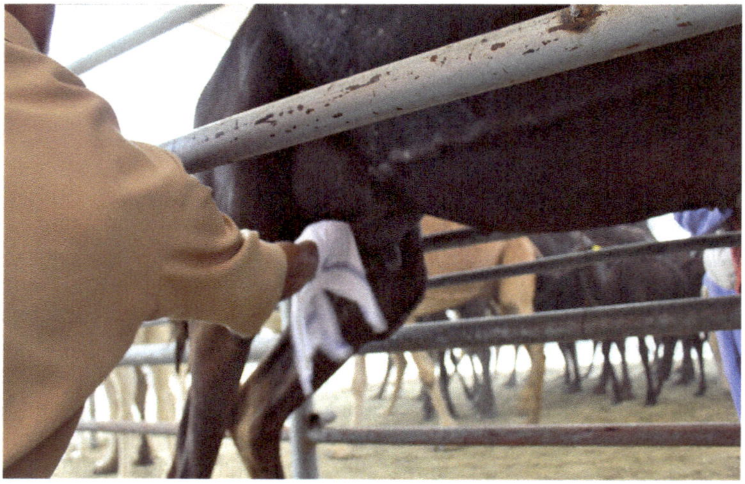

Fig. 5.24 Cleaning the udder with a washable individual wipe

second dedicated box. The used wipes are washed between each milking. If the floor is not clean, it is preferable not to put the equipment directly on the floor, but on a high place. Finally, it is useful to prevent the camels from sitting immediately after milking to give the sphincter time to close properly,

– *Hygiene of the raw matter*: when milking by hand, it is advisable to eliminate the first milk streams, which are more often potentially contaminated, to avoid dipping the finger in the milk to lubricate the teats and to cover the containers containing milk if it is not in closed containers. Milk should be kept at as moderate temperature as possible during storage and transport (put cans in the shade or in a humidified place) or placed quickly in a refrigerated tank. The waiting time between collection and cold storage (4 °C) should be as short as possible, as should the transport time if the cold chain is difficult to maintain. Finally, it is essential to remove milk from camels with clinical mastitis and to treat mastitis, as well as to respect the waiting period for antibiotics if they are treated. Regular detection of subclinical mastitis with, e.g., a California Mastitis Test (CMT, Box 5.1), is also recommended.

Box 5.1 Teepol or CMT (California Mastitis Test) (Fig. 5.25)
The California Mastitis Test or CMT is a quick and efficient test that indirectly measures the level of white blood cells (leukocytes), which indicate the presence of pathogenic bacteria in the milk, as leukocytes are the udder's defense system. The CMT reagent, in the presence of leukocytes, causes an agglutination reaction, the more marked the leukocyte level. This test is regularly used in dairy camels on large dairy farms (Fig. 5.25). The test material consists of (i) a special tray or glass and (ii) a reagent bottle.

The procedure is simple. It consists of the following steps: (i) remove the first few streams of milk from each teat; (ii) collect a few streams of milk in the tray or glass and add about the same amount of reagent; (iii) give a circular motion to the tray or glass to mix the reagent and milk well; then (iv) read the reaction, i.e., the presence or absence of lumps indicating the agglutination reaction (Table 5.1); and (v) assess the pH with a colored indicator (bromocresol purple), which turns purple when the pH is close to 7, indicating the presence of mastitis. After use, the tray is rinsed with clean water, without lye, to avoid false-positive reactions later.

This set of rules constitutes the "good milking practices," which ensure the production of milk of good hygienic quality. The impact of milking practices on the quantitative importance of bacterial contamination has been extensively verified (Fig. 5.26).

In the examples in Fig. 5.13, the bacterial contaminations assessed from coliform counts and total aerobic mesophilic flora (TAMF) were as follows:

(a) Low hygiene level: four million coliforms/ml and 500,000 FMAT/ml of milk
(b) Average hygiene level: one million coliforms/ml and 350,000 FMAT/ml of milk
(c) High hygiene level: 1000 coliforms/ml and 18,000 FMAT/ml of milk

Milk quality control by somatic cell count (SCC) is practiced in large modernized production units such as the Camelicious farm in Dubai but is by no means systematic (Nagy et al. 2013). Overall, standards close to those used for cow's milk can be applied (target: less than 300,000 cells/ml).

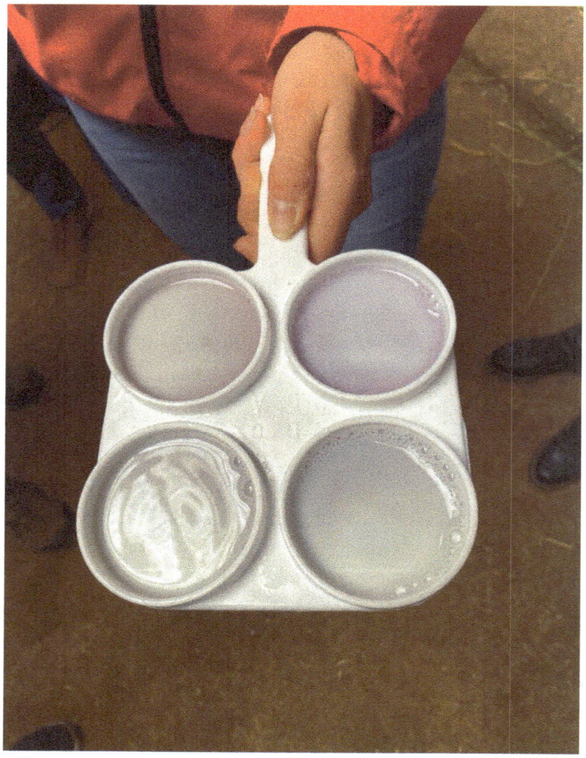

Fig. 5.25 CMT tray with four cups corresponding to each udder quarter. The presence of purplish lumps indicates that the quarter is affected. The reading is done according to the grid reported in Table 5.1

5.3 Rearing Young

5.3.1 Youth Sustainability

The survival and good growth of the young is an essential factor for the numerical productivity of the herd, the quality of the replacement females, and the value of the young males for meat production or breeding. The viability of the calf depends to a large extent on the care given immediately after birth (see Sect. 5.1.3) and on the hygienic conditions during the first 6 months of the calf's life, the most critical period. Indeed, 80% of all mortality in young camels occurs in the first 6 months of lactation (Fig. 5.27).

The causes of this mortality have multiple origins. It can be due to infectious disorders affecting the digestive system (diarrhea is the most frequent disease in the young camel) or the respiratory system, especially in cold periods. But the camel is also susceptible to plant poisoning, accidents, gastrointestinal parasitism, and malnutrition (see Chap. 6 on diseases). The survival of the young animal is as much a question of husbandry practices (colostrum distribution, weaning, and suckling

Table 5.1 Guide to the interpretation of a CMT reaction in dairy camels

Reading aspect	Interpretation		
	Rating	Individual milk	Mixed milk
No reaction. Liquid consistency, gray color	0	Normal milk	Between 0% and 20% of positive camels in the herd
Slight gel flakes that disappear after 10 seconds (purplish gray color)	1	Latent mastitis. Subclinical mastitis Irritating milking	Approximately 30% of camels are positive
Slight persistent frost in the form of lumpy filaments (grayish purple color)	2	Subclinical mastitis Irritating milking	About 40% of camels are positive
Thick gel adheres to the bottom of the cup in a viscous mass when the plate is rotated	3	Well-established mastitis	Approximately 60% of camels are positive
Gel with the consistency of egg white (dark purple color)	4	Well-established mastitis	Approximately 80% of camels are positive

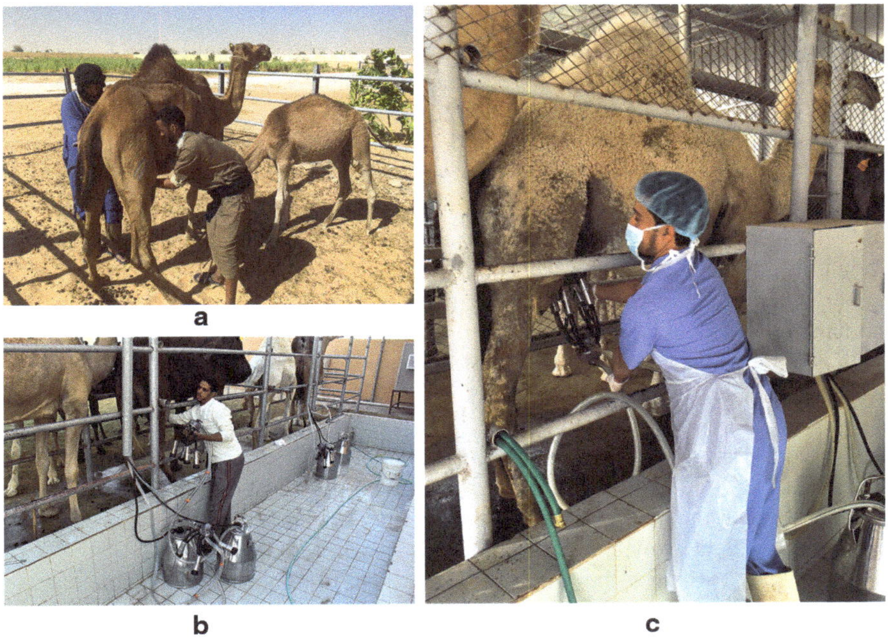

Fig. 5.26 Three levels of hygiene in milking practices. (**a**) Low level; (**b**) medium level; (**c**) high level

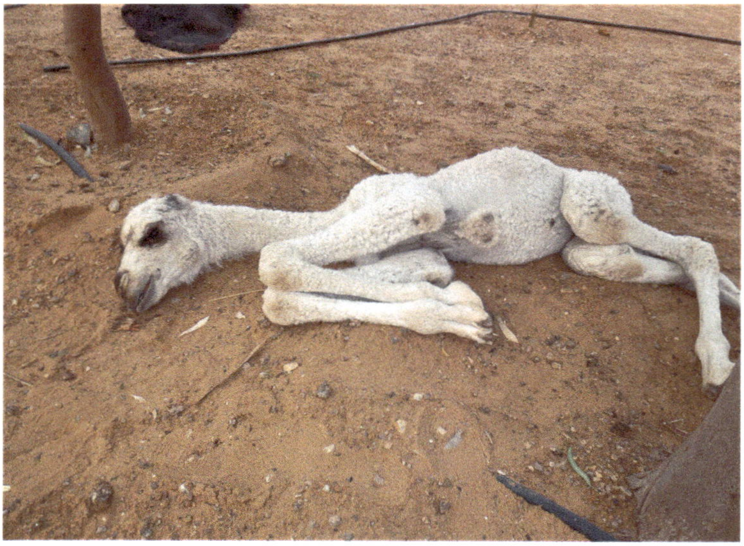

Fig. 5.27 Mortality of the camel occurs in 80% of cases before 6 months. The first month is the most critical

practices) as it is of the immune status of the young animals. Some consider that the criteria for adaptation to a hostile environment imply an urgent natural selection leading to the survival of only those individuals able to face the conditions of a difficult environment. However, the more intensive dairy systems constitute more artificial environments where these criteria are less at stake. It can therefore be considered that breeding practices are more decisive in this context, especially since the causes of mortality are often multifactorial. Moreover, there is a high variability in inter-herd performance, which leads us to believe that improvements can be obtained by improving breeding practices, particularly in the care of young animals.

In addition to the care listed above after parturition, other protective or preventive actions can be implemented, including the following:

- Ensure the milk is taken properly by the nonweaned calf and that the bottle is disinfected if the baby camel is fed by artificial suckling (Fig. 5.28)
- Avoid letting the young suckle its mother in case of mastitis or edema
- Monitor weight growth to detect potential growth accidents
- Isolate sick camels at the first sign of diarrhea, respiratory, or parasitic disease (ringworm in particular)
- Manage a gradual weaning and therefore be careful when switching from an entirely milk-based diet to a forage-based diet
- Treat the area for ticks

Fig. 5.28 Thoroughly disinfect the bottle if you are feeding by artificial suckling

– Check interactions between camels of the same age group (competition for food, aggressiveness of congeners)
– Carry out serological tests in the event of apparent epidemics, which may be of viral or bacterial origin

5.3.2 Weaning

The transition from an entirely milk-based diet to a diet based on fodder and concentrates is a sensitive period for livestock. In the traditional environment, weaning is delayed as long as the camel is present during milking. This practice is justified in a context of scarce resources. Thus, weaning often takes place around 6 months, sometimes a year or more. In an intensive environment for milk production, late weaning is much less justified, especially because the camel takes a significant part of the milk produced (about 40–75%). It is therefore economically necessary to remove the calf as soon as possible without risking its survival and growth. Several studies tend to show that the camel can tolerate the early use of a forage-based feed ration very well. Moreover, there is little weaning stress as in many other species. Early weaning in the traditional environment rarely occurs before 3 months of age and relies on various coercive methods, such as placing a piece of wood in the calf's nostrils to provoke a defensive reaction from the mother. The most common technique, especially among Arab camel breeders, is the use of a *chmel*, a piece of cloth or vegetal weave that prevents the calf from entering the

Fig. 5.29 (a) Use of the *chmel* to protect the udder and prevent suckling outside of milking time. (Photo S. Saleh); (b) cleaning the *chmel* on a Saudi farm

udder between milking times. However, such a device is a real nest for microbes if it is not regularly cleaned, which is why it should be prohibited in more intensive dairy systems (Fig. 5.29).

In intensive dairy farms, weaning can be started as early as 6–7 weeks after calving depending on the growth of the camel. After this time, the calf is separated from its mother during the day, and the she-camel starts to be milked in the morning. This transitional period lasts 2 or 3 weeks, during which the camel gets used to eating fodder and a little concentrate, which allows the development of its forestomachs. At the end of this period, the calf is completely separated from its mother and is socialized in the cohort of young born in the same season than it. However, the camel can be left with its mother for 30 minutes after milking so that it can benefit from the residual milk until 5–6 months. After this period, the presence of the calf is no longer necessary.

5.3.3 Growth Control and Measurements

The average birth weight of a camel is 35 kg with a range of 26–51 kg. Below 25 kg, the chances of survival are very low. This weight would depend more on intrauterine growth (thus maternal genetics) and the environment, especially the mother's diet at the end of gestation, than on a possible hereditary genetic potential (heritability appears low). This low heritability could be associated with the absence of selection on this criterion. Growth control is an important element of animal monitoring. Overall, this growth follows a classic S curve but with performances that vary greatly depending on the production system.

Under traditional conditions, camel weight growth is in the range of 190–310 g/day during the first year. Under more intensive conditions, the average daily gain (ADG) varies from 440 to 580 g. Higher values are sometimes reported in the literature (up to 750–800 g/d). In general, growth begins to decline after 6 months (about 500 g/d in intensive farms), reaching about 200 g/d after 2 years. Growth lasts

Table 5.2 Comparison of some prediction equations for live weight (LW) from height at withers (HG), chest circumference (TP), and abdominal circumference including hump (TA) for a real average weight of 406 kg

Equation	Estimated weight	Author	Location	Date
LW= 53 * (HG * TP * TA)	404	Boué (1949)	Algeria	1949
LW= 52 * (HG * TP * TA)	397	Graber (1966)	Chad	1966
LW = 507 * TP – 457	536	Wilson (1978)	Sudan	1978
$LW= 6.46 * 10^{-7} * (HW + CG + HG)^{3.17}$	399	Field (1984)	Kenya	1979
LW= 3.06 * TP – 209.6	454	Bucci et al. (1984)	Egypt	1984
LW = 50 * (HG * TP * TA)	382	Yagil (1994)	Israel	1994

Source: Boujenane (2019)

until the age of 7 years, but it is the first 12 months that are essential to control. There are two methods for this: either barometric measurements or weighing. There are several equations for estimating weight from a few measurements, but these equations were developed on adult animals (see Table 5.2) and are not necessarily suitable for young animals. However, it is possible to use a few measurements that are relevant for growth (height at withers, thoracic circumference) without trying to estimate weight. However, the use of a scale is more reliable if the weighing is always done under the same conditions, for example, in the morning on an empty stomach and before any watering and with equipment that is suitable for weighing animals.

There are several weighing systems available (Fig. 5.30). In field surveys or for weighing very young camels only, a removable weighing system with a scale may be sufficient. In large dairy units, it is possible to install a weighing system with an electronic scale, but such a system requires more infrastructure with a corridor and appropriate fencing.

Weight assessment of adult animals in the absence of a weighing system using equations should also be treated with caution. The most common barometric measurements include thoracic girth, height at withers, and abdominal circumference. A recent comparative analysis of six equations published in the scientific literature (Table 5.2) showed that, in relation to the actual weight (on average 406 kg), the average estimated values varied between 382 and 536 kg, the best predictions being those proposed by Field (1984) in Kenya or Boué (1949) in Algeria.

Schwartz and Dioli (1992) also proposed a live weight prediction equation taking into account the hump, PV (in kg) = HG*TP*HB*50, where HB is the height of the hump (in cm).

A single equation has been proposed to estimate live weight of growing animals as a function of age based on other measurements, namely, neck circumference (TC), thigh circumference (CC), and hump length (LB) according to the following formula (Kamili et al. 2006):

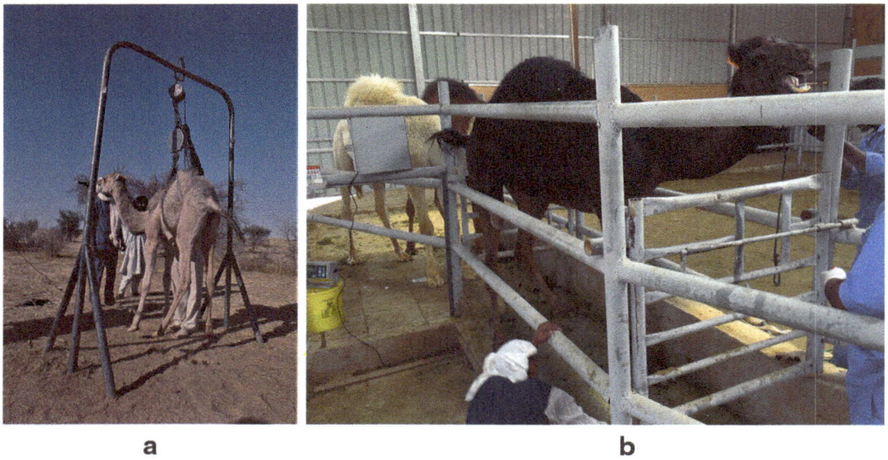

Fig. 5.30 (**a**) Demountable hanging scale system with weighing in Niger; (**b**) weighing system with electronic scale in Saudi Arabia

$$\text{PV (kg)} = 4.06^{*}\ \text{age (years)} + 3.05^{*}\ \text{TC (cm)} + 3.38^{*}\ \text{CC (cm)} + 1.38^{*}\ \text{LB (cm)} - 191$$

Many other measurements have been proposed, but less to estimate the live weight of the animal than to determine the morphology of the ecotypes in order to distinguish camel phenotypes: length of the neck and head, length of the body, abdominal circumference behind the hump, length of the limbs, sub-sternal height, etc. (Fig. 5.31).

5.4 Feeding Management

Good feed management is essential to ensure the maintenance and production performance of farm animals. This management is based on the knowledge of the needs according to the physiological status of the animal (gestation, lactation, growth) and the available food resources in order to determine the optimal rations.

5.4.1 Feed Requirements

Although the nutritional requirements of the camel are lower than those of other species, due to more efficient digestive use and recycling mechanisms (see Sect. 4. 2.1), it is necessary for the camel to eat and drink in proportion to its size and production. The feed requirements concern the necessary supply of proteins, energy, minerals, and water for maintenance and production. Therefore, they vary greatly,

Fig. 5.31 Measurement of udder circumference of an adult camel in Saudi Arabia

depending on what is expected from the animal. Furthermore, the gradual shift toward intensified farming systems for milk production is accompanied by a considerable change in the feeding system. Free-grazing in pastoral areas with low nutrient resources but high diversity (see Sect. 4.2.3) is being replaced by monotonous rations with high energy and protein concentration, without clear evaluation of the effects on productivity and metabolic disorders. It is therefore essential to provide a balanced diet that stimulates the expected milk production.

5.4.1.1 Energy Requirements

It is generally considered that the maintenance requirements of adult dromedaries are in the order of 1.2 forage units (FU) per 100 kg live weight. One FU corresponds to the energy value of 1 kg of barley harvested at the mature grain stage. A distinction is also made between fodder units for milk production (FUL) and fodder units for meat production (FUM). The FUL is the amount of net energy that can be absorbed during lactation or maintenance of the ruminant and corresponds to 1700 kcal for 1 FUL. The FUM is the amount of net energy that can be absorbed during fattening if the DWG is greater than 1000 g/d, i.e., 1820 kcal for 1 FUM. In the Anglo-Saxon system, energy is measured in megajoules (MJ) rather than in FU. One FUL corresponds to 7.1 MJ. For animals on daily movement, the maintenance energy requirements are increased by 20% to account for muscle expenditure. The energy requirement of a dairy camel is obviously dependent on the amount of milk she produces.

The following summary table shows the energy requirements according to the status of the animal (Table 5.3), bearing in mind that the range of values depends on the weight of the animal. On average, the energy requirement to produce one liter of milk is 5 MJ or 0.7 FUL. A camel producing ten liters of milk will therefore need an additional 7 FUL.

5.4.1.2 Nitrogen Requirements

In Sect. 4.2.1, we saw that protein and nonprotein nitrogen in forages is metabolized to ammonia, from which the microbes in the C1 compartment synthesize the digested protein in the gut. However, a part of proteins in the ration are also absorbed in the gut. These two sources of protein make it possible to cover the animal's maintenance requirements, which are estimated at 90 g of crude protein per 100 kg of body weight. This corresponds to lower values than those of other domestic herbivores, due to the urea recycling capacities already mentioned above. Milk production requires about 55 g of digestible protein per kilogram of milk. However, little is known about the distribution of dietary protein into milk and muscle protein in large camelids, but despite urea recycling capabilities, dromedaries waste a large proportion of ingested protein, and only 15% of what is ingested is retained in the body. The requirements according to the status of the animal are summarized in Table 5.3. These requirements are significantly lower than those of a cow with the same weight.

5.4.1.3 Water Requirements

Given the ability of large camelids to resist thirst and specially to save water, water requirements are quite low compared to other livestock species. It is estimated that under unfavorable climatic conditions (heat and drought), daily requirements are around 6 l per 100 kg of body weight. This is reduced to half under favorable conditions (cool, wet season). In lactating females, milk production increases water losses due to the water content of the milk and the increase in basic metabolism induced by lactogenesis. The production of 1 liter of milk requires an additional 1.5 l of water. During the last third of gestation, pregnant females increase their water requirement by about 20%.

Table 5.3 Daily energy and nitrogen requirements according to the physiological status of camels for weights ranging from 100 to 600 kg (growth) and milk production ranging from 1 to 10 l/d of milk

Requirements/ day	Post-weaning growth	Nonpregnant females	Pregnant females	Lactating females
Energy (UFL)	2–7	3.5–5.1	3.9–8.1	5.6–12.9
Energy (MJ)	14.6–54	24.9–42	28.1–57.6	40–92
Protein (g)	195–495	144–321	234–492	470–608
Water (l)	6–18	18–36	25–45	25–45

[a]For a growth of 500 g/d

5.4.1.4 Mineral and Vitamin Requirements

Salt (sodium chloride) requirements are high. They are estimated at 20 g per 100 kg of body weight. For a 400-kg animal, this corresponds to an intake of 28–34 kg of salt per year. Above these values, urinary salt excretion increases in proportion to the amount ingested. Given the salt content of milk, the lactating camel's requirement is around 2.5 g/l of milk. Gestation, however, does not affect the requirement. Salt supplementation can be provided by salt blocks, especially if the animals do not have access to halophyte plants. The digestive utilization coefficient of phosphorus and calcium is higher than in cattle, which means that large camelids absorb these major minerals better (Faye et al. 1999). The maintenance requirement for camels is estimated at 4-g calcium and 2.5-g phosphorus per 100 kg of body weight. Based on milk content, the requirements of the lactating camel are estimated at 1.9-g calcium and 1.1-g phosphorus per liter of milk produced. The intake required for fetal bone development during the last third of gestation is not well known. By extension of the results observed in other species, we can however estimate these needs at 9.5 g of calcium and 5.5 g of phosphorus per day. The growth requirements are however not well known. Regarding magnesium, there are no precise references. However, it is known that a daily intake of 3 g per 100 kg of body weight is sufficient to maintain normal blood levels.

Regarding trace elements, the needs are fairly well known. The regulation of copper is comparable to that of cattle, and the requirements for camelids are the same, i.e., about 15 mg per 100 kg of body weight. On the other hand, the metabolism of zinc is different from that of cattle, and if the recommended intake is about 60 mg per 100 kg of body weight, i.e., a value comparable to that of cattle, it is likely that the real needs are lower. The same is true for selenium, as the dromedary is very sensitive to supplementation (Bengoumi et al. 1998). The recommended intake is twice as low as for cattle, i.e., 0.06 mg per 100 kg of body weight. Because camel milk is rich in iron and fodder in arid and semiarid areas contains sufficient absorbable iron, there is no need to supplement dromedaries or camels with this element. Table 5.4 shows the requirements for these main mineral elements.

Vitamin requirements are not known. Given the exceptional richness of camel milk in vitamin C (ascorbic acid), the needs are undoubtedly important in lactating camels, but because ascorbic acid is synthesized in the liver, as in ruminants, there is no need for specific supplementation. On the other hand, supplementation with fat-soluble vitamins (particularly vitamins A and E) may be considered due to their beneficial effect on reproduction. Concentrations of β-carotene (provitamin A) are very low in both blood and milk, which explains the translucent color of dromedary plasma and the absence of cream color in milk. The plasma content of vitamin E is also comparatively lower than in other domestic herbivores. In contrast, folic acid (vitamin B9) and biotin (vitamin B7) had similar plasma concentrations to those of cattle. However, these results are insufficient to specify requirements, and vitamin deficiencies are not well known in this species, although there is probably a link between mange susceptibility and vitamin A deficiency. On the other hand, the water-soluble vitamins of the B group are either in equal or lower quantities, especially in milk, compared to cow's milk, for example. But deficiencies are

Table 5.4 Daily mineral requirements according to the physiological status of camels for weights ranging from 100 to 600 kg

Element	Growth	Not pregnant	Gestating	Lactating
Sodium (g/d)	20–60	60–120	60–120	72–132
Calcium (g/d)	4–12	12–24	21–33	21–33
Phosphorus (g/d)	1.8–5.4	5.4–11	10–16	10–16
Magnesium (g/d)	3.9	9–18	?	?
Copper (mg/d)	15–45	45–90	45–90	45–90
Zinc (mg/d)	60–180	180–360	180–360	180–360
Selenium (mg/d)	0.12–0.36	0.36–0.72	0.36–0.72	0.36–0.72

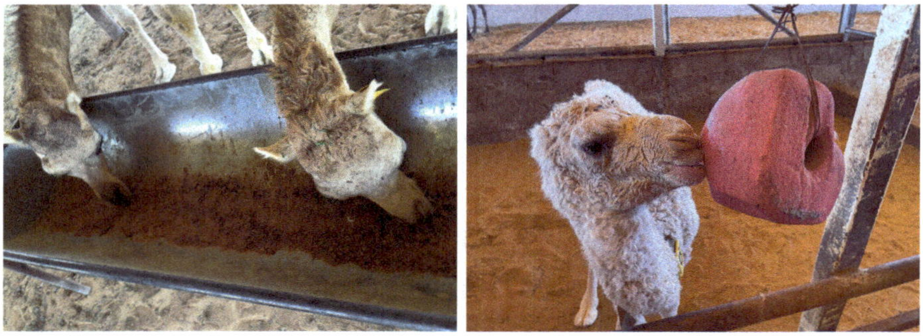

a b

Fig. 5.32 (**a**) Distribution of mineral and vitamin supplements in the form of crushed blocks to allow better ingestion. (**b**) Salt lick distribution to camel calf

never described in the literature. As camelids normally live in very sunny regions, the synthesis of vitamin D by the skin under the effect of the sun appears to be very high (the level in plasma is 10 times higher than in cow's milk). However, the question could arise in farms in northern countries. In general, minerals and vitamins are provided by licking blocks enriched with trace elements and vitamins. However, unlike cattle, dromedaries and Bactrians do not "lick" these blocks but rather tend to bite them. It is therefore advisable to provide the mineral and vitamin supplements in the form of powder or crushed blocks (Fig. 5.32).

5.4.2 Potential Food Resources

As animals living in arid and semiarid zones, large camelids are accustomed to the vegetation of dry zones. Plants likely to be consumed by the dromedary in natural conditions, knowing the diversity of pastoral situations from Mauritania to China, are several and have widely varied nutritional values. However, there are some particularities of these plants that can be summarized as follows:

- They are often small plants, with a strong root system, equipped with devices to limit evapotranspiration (reduced, acicular, varnished, or fatty leaves)
- Herbaceous plants are therophytes whose seeds have a long-lasting dormancy and germination power that can be preserved for a long time
- Perennial plants have the ability to survive in a slow-moving state for long periods of time and have efficient root absorption and water retention mechanisms
- The rhizosphere (i.e., the underground part) of these plants is developed, which leads to a great dispersion of the aerial tufts, often associated, moreover, for the herbaceous plants, with great mobility

In pastoral regions where large camelids live, the floristic richness is greater than one would expect, even if each species is represented by a relatively small number of individuals. However, the dispersion of this vegetation cover results in low pastoral productivity per unit area. The animal load per hectare is therefore generally very low, and it is estimated that more than 20 ha per year is needed to feed an adult camel. This is, of course, an average, and there is great variability between regions and seasons (Fig. 5.33).

Another important element of camel feed resources is the use of woody resources, which may be superior to herbaceous resources at the desert margins. The nutritional value of "aerial pastures" is often higher than that of grasses and other herbaceous plants. In addition, the length of the animal's neck allows it easy access to the foliage of good-sized trees (see Sect. 4.2.3).

In intensified systems, camelids may have access to other resources from agriculture (broken rice or wheat bran, groundnut tops, barley, brewers' grains, oil by-products such as olive cake, cotton, or groundnut cake, etc.) or commercial concentrates to supplement a basic ration that may come from rangeland, alfalfa hay, or other fodder crops from irrigated areas (Fig. 5.34). Silage trials have been carried out using grasses (notably Rhodes grass – *Cenchrus ciliaris*). Rations based on urea-treated straw (to increase the proportion of non-protein nitrogen) or

a b

Fig. 5.33 (**a**) Camels on the rangelands in the Mathrouh region of Egypt; (**b**) bactrian camels and hybrids on rangelands in the Akshi region of Kazakhstan

Fig. 5.34 Feeding alfalfa hay from irrigated fields to dairy camels in Saudi Arabia

molasses-urea blocks sometimes proposed as a supplement for dairy cows should be avoided because of the risk of hyperuremia.

In northern countries, feeding conditions based on temperate grasslands can be particularly rich, especially in spring, and can cause various metabolic disorders. It is therefore advisable to add ballast (e.g., wheat straw) at a rate of at least 1 kg/day.

5.4.3 Calculation of Ration

Energy is provided in the feed ration either by the celluloses and hemicelluloses constituting the cell walls of the plants consumed or by the starches and complex sugars present in abundance in concentrates such as barley or wheat grains. These nutrients are transformed into simple sugars for the former and glucose for the latter, but in all cases, these sugars are transformed into volatile fatty acids (VFAs): acetate, propionate, and butyrate. These VFAs are rapidly absorbed by the stomach walls and pass into the bloodstream to be metabolized in the different organs as energy. With rations rich in cellulose, the acetate content is higher, whereas with rations richer in concentrates, propionates and butyrates predominate, as well as the amount of glucose absorbed in the intestine, especially with grain-rich feeds. However, propionates and butyrates are products preferentially used by the adipose tissue. In lactating females, body deposits therefore increase at the expense of mammary secretions, unless they are high-producing females. However, in low-producing

Fig. 5.35 Obesity in a camel resulting in excessive fat storage in the hump

camels, as it is the case for most African "breeds," excessive intake of concentrates will not necessarily result in more milk or richer milk, but in filling of the hump.

It is therefore recommended in "intensive" rations to:

- Mix concentrates with fodder, splitting the distribution of concentrates
- Limit the intake of easily fermentable carbohydrates (starch, date refusals)
- Not destroy the fibrous structure of fodder and cereal husks (avoid chopping straw and hay)
- Limit the input of concentrates to a maximum of 40–45% of dry matter
- Limit the FUL energy content of concentrates to 60%, i.e., an energy density of between 0.91 and 1.17 FUL/kg DM
- Prefer barley or wheat as concentrates rather than maize or sorghum to maintain a ratio (acetates + butyrates)/propionates of about 3.5.

These changes may lead to a preferential use of the end products of digestion by the udder rather than for fat reserves. The addition of concentrates should therefore be done with caution, especially because this addition does not have a negative effect on forage digestion. For example, the addition of concentrates at a rate of 20% of the basic ration increases intake by 17–22%, which is explained by a greater speed of elimination of the products of microbial digestion (volatile fatty acids, ammonia) linked to a more rapid turnover of the liquid phase of digestates and a greater absorption by the digestive wall. As a result, rations rich in concentrates leading to an excess of propionates over acetates can induce obesity (Fig. 5.35).

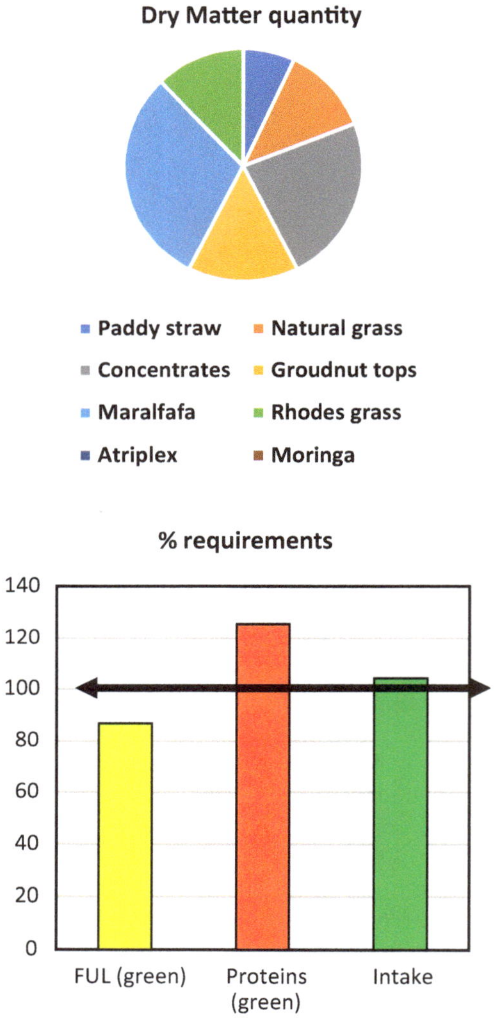

Fig. 5.36 Simplified ration calculation sheet for a dairy camel

When the forage is rich in nitrogen, total consumption increases significantly, while poor products, such as straw, are ingested in smaller quantities. However, this decrease is compensated by the good valorization of poor rations. Indeed, the digestibility of straw is about 5% higher than that measured in sheep, and even reaches 7% when dehydrated.

There is no standard ration for intensive dairy farming, as it depends on local availability. However, a ration can be established from available fodder and agricultural by-products in order to cover as closely as possible the energy and protein requirements according to the expected milk production. In Fig. 5.36, for example, a spreadsheet is presented based on the energy and protein value of some feeds in the ration for a production of 8 liters of milk by a 450-kg camel ingesting 2.5-kg

DM/100 kg PV. The ration presented here largely covers the protein requirements but only 90% of the energy. It is therefore desirable to rebalance the ration by adding more energy while reducing protein.

References

Atigui M, Hammadi M, Barmat A, Farhat M, Khorchani T, Marnet PG (2014) First description of milk flow traits in Tunisian dairy dromedary camels under an intensive farming system. J Dairy Res 81:173–182. https://doi.org/10.1017/S0022029914000089

Ayadi M, Musaad A, Aljumaah R, Matar A, Faye B (2016) Effects of manual udder stimulation on milk partitioning and flow traits during the machine milking in dairy camels. J Camel Pract Res 23(1):1–5

Ayadi M, Musaad A, Aljumaah R, Matar A, Konuspayeva G, Abdelrahman M, Abid I, Bengoumi M, Faye B (2018) Machine milking parameters for an efficient and healthy milking in dairy camels (Camelus dromedarius). J Camel Pract Res 25(1):81–88

Bengoumi M, Essamadi AK, Tressol JC, Chacornac JP, Faye B (1998) Comparative effect of selenium concentration and erythrocyte gluthatione peroxidase activity in cattle and camels. Anim Sci 67:461–466

Boué A (1949) Essai de barymétrie chez le dromadaire Nord-africain. Rev élev Méd Vét Pays trop 3(1):13–16. https://doi.org/10.19182/remvt.6857

Boujenane I (2019) Comparison of body weight estimation equations for camels (*Camelus dromedarius*). Trop Anim Health Prod 51(4):1003–1007. https://doi.org/10.1007/s11250-018-1771-8

Bucci TJ, Soliman AM, Botros BAM, Kerkor ME (1984) Abdominal circumference at the hump as an index of body weight in dromedary camels. Indian Vet J 61:26–30

Faye B, Bengoumi M, Tressol JC (1999) Comparative trace-element excretion in camel and cow. J Camel Res Pract 6:19–25

Field CR (1984) Camel growth and milk production in Marsabit district, Northern Kenya. In: Cockrill WR (ed). The camelid an all-purpose animal. Proc. Khartoum Workshop on Camels, December 1979, Uppsala, Scandinavian Institute of African Studies, 209–230

Graber M (1966) Etudes dans certaines conditions africaines de l'action antiparasitaire du Thiabendazole sur divers helminthes des animaux domestiques. Rev Elev Méd Vét Pays Trop 19:527–543. https://doi.org/10.19182/remvt.73188

Kamili A, Bengoumi M, Faye B (2006) Assessment of body condition and body composition in camel by barymetric measurements. J Camel Pract Res 13(1):67–72

Konuspayeva G, Faye B, Loiseau G, Levieux D (2007) Lactoferrin and Immunoglobin content in camel milk from Kazakhstan. J Dairy Sci 90:38. https://doi.org/10.3168/jds.S0022-0302(07)72606-1

Nagy P, Faye B, Marko O, Thomas S, Wernery U, Juhasz J (2013) Microbiological quality and somatic cell count in bulk milk of dromedary camels (Camelus dromedarius): descriptive statistics, correlations, and factors of variation. J Dairy Sci 96:5625–5640

Nagy P, Faigl V, Reiczigel J, Juhasz J (2015) Effect of pregnancy and embryonic mortality on milk production in dromedary camels (*Camelus dromedarius*). J Dairy Sci 98:975–986. https://doi.org/10.3168/jds.2014-8546

Schwartz H, Dioli M (1992) The one-humped camel in Eastern-Africa. Verlag, Weikersheim, p 282

Tibary A, Anouassi A (1997) Theriogenology in camelidae, Actes éditions IAV Hassan II, Rabat, Maroc, 489 p

Wilson RT (1978) Studies on the livestock of southern Darfur, Sudan. V. Notes on camels. Trop Anim Health Prod 10:19–25. [en ligne]. https://doi.org/10.1007/BF02235296

Yagil R (1994) The camel in today's world. In: A handbook for camel breeding. Deutsche Welthungerhilfe, Bonn, p 73

Animal Health Management and Veterinary Practices in a Camel Farm

6

Abstract

Disease control is an important item including good techniques of restraining, sedative and anesthesia practices, clinical examination and sampling, laboratory analysis, treatment of the sick animals, prevention, and necropsy of dead animals, most of these items being under veterinary control. The main diseases affecting the different organs and anatomical parts of camel are described, and possible treatment and prevention are suggested. It includes skin diseases, eyes and sinus conditions, locomotion disorders, digestive and respiratory syndromes, reproduction diseases, and some specific diseases as mange or trypanosomosis.

Keywords

Health management · Veterinary practices · Constraints · Necropsy · Semeiology

The breeding of large camelids is considered as a risky breeding because of the slow life cycle and the relatively low survival rate of the young. Good sanitary management is therefore essential to optimize productivity, especially since the symptomatology in this species (i.e., the expression of symptoms) is often frustrated and, therefore, the diagnosis difficult. It is therefore necessary to favor sanitary management over piecemeal care. Overall, large camelids are not very susceptible to the major infectious diseases that affect livestock in tropical areas, such as foot-and-mouth disease, rinderpest, or peripneumonia (PPCB). However, this resistance should not obscure the heavy toll paid to other widespread diseases. Furthermore, the intensification of breeding systems and the adaptation of the species to temperate ecosystems have led to a new configuration of health risks. In this section, we will successively address the issue of restraint, the elements to be known for the clinical examination of animals, some elements on the main diseases encountered, and some data on the interpretation of blood tests.

B. Faye et al., *Large Camel Farming*, https://doi.org/10.1007/978-94-024-2237-5_6

6.1 Restraint Techniques

Camelids are large animals, whose clinical examination or various manipulations for treatment or reproductive monitoring require restraint in order to avoid violent or aggressive reactions by the operators while limiting stress and the risk of accidents for the animal and themselves. Although generally placid in character (especially dairy camels that are used to being handled), the animal is nevertheless capable of violent gestures (biting, kicking, rejection of ruminal contents, crushing under its sternal plate as it would do to subdue a fellow animal in a fight) and unexpected reactions caused by severe stress. In addition, it vocalizes its discontent, anger, or fear, which can be very destabilizing for the operator. The implementation of good restraint techniques is therefore essential and often requires several operators.

6.1.1 Head Restraint

The head of large camelids represents a significant mass (15–20 kg) planted at the end of a very long neck of more than 1 m. Such an anatomical arrangement constitutes a mass endowed with a strong inertia capable of knocking out any operator having to intervene near the head (e.g., for a blood collection at the jugular). Moreover, the animal is capable of biting deeply and crushing by lateral shearing of the jaws. The male's canine teeth can cause particularly deep and debilitating wounds (Fig. 6.1). Good head restraint is therefore of prime importance.

If a large camelid is to be handled frequently, it is advisable to always use a halter. There are several types of halters, most of which corresponding to those used for horses (Shetland halter for young camels or cob size halter for adults). In the Gulf countries, special camel halters are also available in *camel souks* (Fig. 6.2).

The use of a simple rope placed around the neck is not recommended (risk of cervical dislocation, laceration, strangulation), except to inflate the jugular vein when taking blood.

Muzzles are used to prevent biting and rumen juice rejection, but in general they are used mainly in racing camels to prevent the consumption of unwanted forage by the owner, with the muzzle being removed when the appropriate feed ration for this type of animal is fed. There are also several types (Fig. 6.3).

In the absence of a halter and/or muzzle, a lace can be used to effectively close the jaw, leaving the mucous membranes accessible for examination. Alternatively, a rope can be placed around the lower jaw and tightened to contain the head and allow observation of the mouth (Fig. 6.4).

Holding the head, for example, to place an ear tag, can also be done by firmly holding the lower and upper lips (Fig. 6.5). Holding by the ears is possible, but more difficult because the ears of large camelids are small.

Many camel farmers use a nose ring (Fig. 6.6) or a wooden peg that punctures the animals' nasal septum. These devices do not allow for muzzling but are generally used for caravanning or walking to tie animals together.

Fig. 6.1 The male has well-developed canines that can cause debilitating bites if not properly restrained

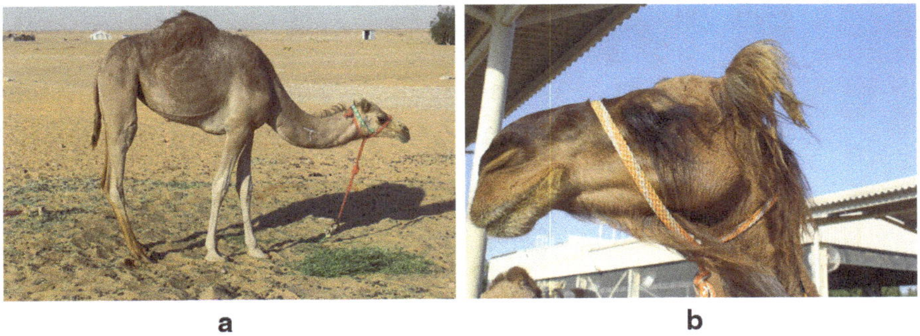

Fig. 6.2 (**a**) Rope halter for racing camel in Saudi Arabia. (**b**) Ribbon halter for dairy camel in Dubai

In any case, they are not very practical for holding the head. On the one hand, they do not prevent biting, and moreover, by gesticulating, the animal can tear off the ring and cut the skin of its nostrils. The peg is stronger, but the installation is more painful and is now discouraged by animal rights groups (Fig. 6.7).

Nasal pegs are also used to prevent camels from suckling their mothers in order to start weaning or on adults with a tendency to suckle their conspecifics, as this is considered a behavioral anomaly (Fig. 6.8).

Fig. 6.3 (a) Cotton braided muzzle for racing camel. (b) Net muzzle for young male

Fig. 6.4 Holding the head with a rope tightly around the lower jaw. (Photo G. Konuspayeva)

In all cases, head support is essential for many manipulations. It is important to (i) place the head as low as possible in a standing animal to prevent it from rearing up and striking the forelegs, (ii) stretch the neck to avoid head strikes, or (iii) bend the neck over the thorax to avoid sledgehammering by neck swinging (Fig. 6.9).

6.1.2 Limb Restraint

The dromedary is able to kick with both front and back legs in all directions. Even though the Bactrian is a little less agile, this makes the large camelids more

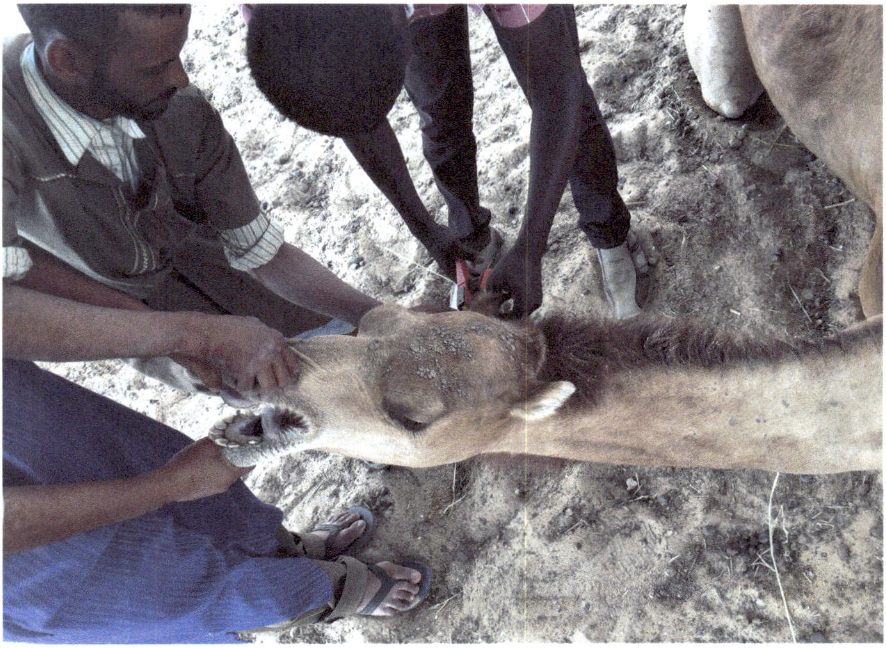

Fig. 6.5 Holding a camel's head by firmly holding the upper and lower lips and extending the neck

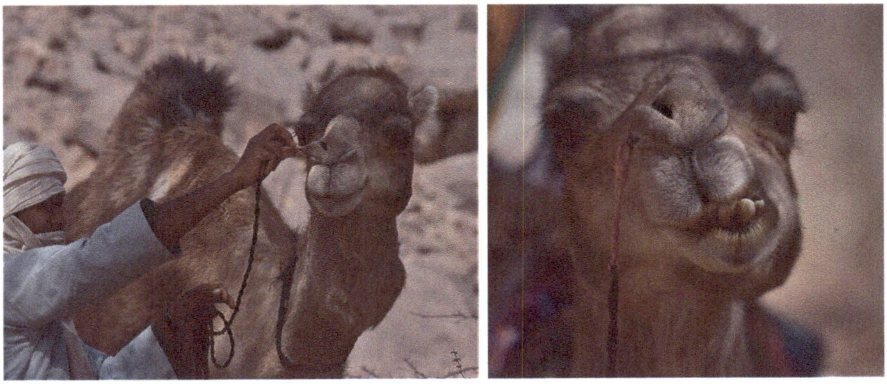

a b

Fig. 6.6 Putting a cord in the nose ring for caravan camels in Algeria

dangerous than other large farm animals. Moreover, being high on its feet and having the attachments on the thorax (front legs) or the abdomen at the fold of the stifle (rear leg) well clear, the range of movement is much greater than in other species. It is therefore preferable to intervene on the animal lying down, because it is then possible to block it in the sternal position by restraining the forelegs and hind

Fig. 6.7 (a) Nasal pinning of a camel in India. (b) Tether to catch the animal and hold it in a caravan. (c) Nasal pegs on a Bactrian camel in Kazakhstan

legs; however, if one must intervene on the standing animal (e.g., on the udder in the female or on the penis in the male), it is preferable to place oneself as close as possible in order to avoid taking the blows at the end of the stroke – capable of breaking the operator's legs – and to hinder the distal part of the limbs.

6.1.2.1 Restraint of the Animal in the Stooped Position

There are several methods of restraining the camel in a kneeling position:

- Holding the forelegs (Fig. 6.10a): pass a rope over the pastern and hold the bent limb by passing over the radius as close to the elbow as possible. Ideally, the rope can then be passed over the neck, and the same is done with the other foreleg (Fig. 6.10b).
- Hind leg support: pass a rope to hold the bent hock, and ideally do it on both hind legs.

Fig. 6.8 Coercive nasal pin to prevent the animal from suckling its conspecifics

Holding the hindquarters and forequarters together allows the animal to be tilted to gain access to the udder for surgical procedures, for example (Fig. 6.11).

When the animal is not trained to lie down, there are several methods to force it. This requires experience with these techniques. One method is to pass a rope behind the hind limbs by two operators on either side of the animal, while a third man bends one of the forelimbs. The first two people pull the rope to push the hind limbs toward the front of the animal, forcing it to bend all its limbs and descend onto its sternal pad. When the animal is braced, it is held in this position by the rope tied around the two hind limbs in addition to the forelimbs. Another technique, which requires at least two experienced operators, is to run around the animal in the opposite direction to each other, wrapping a rope around the limbs and then after two or three turns tightening the grip until the animal tips over (Fig. 6.12).

6.1.2.2 Restraint of the Standing Animal

To restrain the animal in a standing position, the easiest way is to tie the two forelimbs or hind limbs together (there is no need to tie both limbs together on one side, as large camelids walk at an amble). It is best to twist the rope between the two limbs, which makes it stronger and prevents the animal from becoming entangled (Fig. 6.13).

Hind leg restraint is also used to prevent the lactating camel from gesticulating during milking, especially during the period of adaptation to the milking machine. To avoid the risk of kicking, sometimes only one of the front legs is restrained (Fig. 6.14). This technique is also used for animals that tend to wander off at grazing

Fig. 6.9 Neck folded over the chest to avoid head bobbing, a convenient position for ear tags or jugular blood sampling

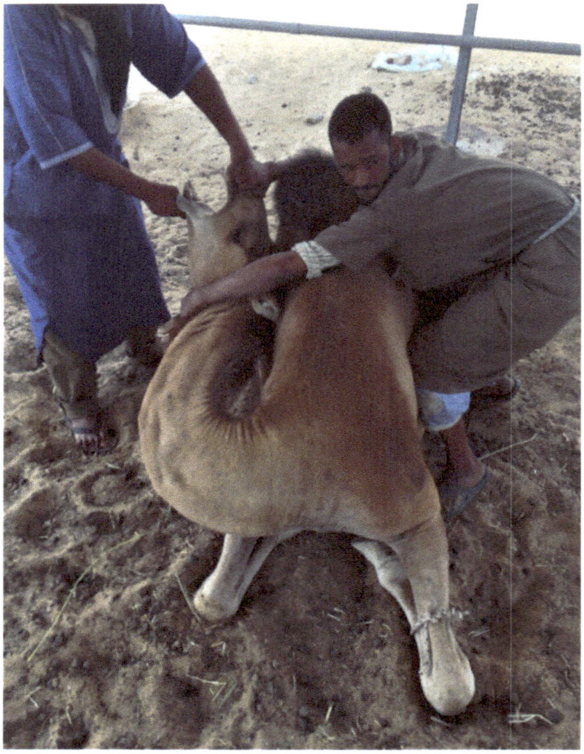

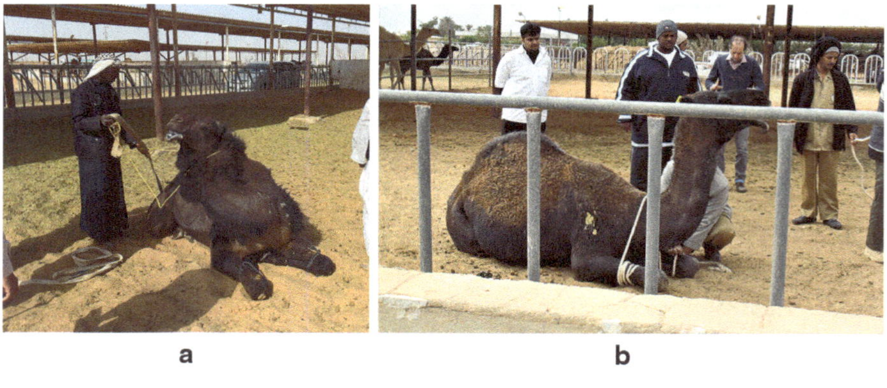

Fig. 6.10 (**a**) Holding both forelegs folded, preventing the animal from standing up. (**b**) Holding the forelegs with the rope passing over the neck. (Photo G. Konuspayeva)

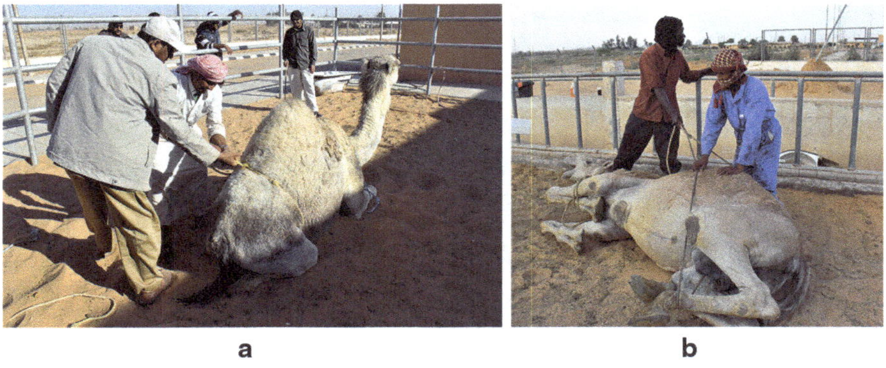

Fig. 6.11 (**a**) Restraint of the hind and front legs. (**b**) Tipping a restrained animal for surgery on an udder affected by a teat lesion. (Photo G. Konuspayeva)

time, thus limiting the radius of their movements. Camelids have very good balance on three legs. This makes it easier to stand still.

In all cases, the tether ropes should be of sufficient diameter not to shear the animal and should be flexible enough to be easily undone.

For individual interventions that require, for example, rectal palpation (reproductive disorders), fecal sampling, urine collection, or even blood sampling, the use of a restraint cage can be useful but requires the animal to be trained (Fig. 6.15a). While it is worth investing in such training for regular interventions on the animal (e.g., monitoring reproduction), getting the animal to enter such a structure at random can be laborious (Fig. 6.15b). Fear or pain may cause the animal to drop to the ground (unlike an equine), and it may become trapped in the cage if it has no side opening.

For interventions on a whole herd, it can be useful to have a corridor where several animals can enter. Round metal structures are preferred and should be sufficiently strong as the strength of a stressed camelid is always underestimated. They should also be sufficiently high (1.60–1.80 m depending on the ecotype for the highest bar), as large camelids, especially dromedaries, can jump (Fig. 6.16). With overly fearful, untrained animals, sedation may be necessary. In general, it is best to have more than one person examining a camelid.

6.2 Sedation and Anesthesia

Sedation consists in allowing the examination of a fearful or untrained animal and thus limiting stress. The protocols are diverse and use a combination of sedatives:

- Xylazine 0.25 mg/kg intravenous (IV) or 0.35–0.45 mg/kg intramuscular (IM)
- Xylazine (0.15 mg/kg) combined with ketamine (2.5 mg/kg) IM
- Xylazine (0.2 mg/kg) combined with butorphanol (0.05 mg/kg) IM

a b c d e f

Fig. 6.12 Restraint of an untrained Bactrian camel in Kazakhstan. (**a**) Placing the rope around the limbs. (**b**) Running around the animal in opposite directions. (**c**) Destabilization of the animal. (**d**) Tipping of the animal. (**e**) Blocking the head. (**f**) The animal is subdued. (Photos G. Konuspayeva)

– Medetomidine (0.01 mg/kg) combined with ketamine (2.5 mg/kg) IM or IV
– Detomidine 0.03–0.06 mg/kg IV

Once the animal has been tranquilized, it must be left in a calm state, and wait for the products to take effect before stimulating it and intervening. Often the animal lies down, and this is the opportunity to seize to restrain it in the sternal position.

Good sedation with local anesthesia allows routine procedures such as taking blood, placing a venous or subpalpebral catheter, paracentesis, palpation, and mouth examination (Fig. 6.17).

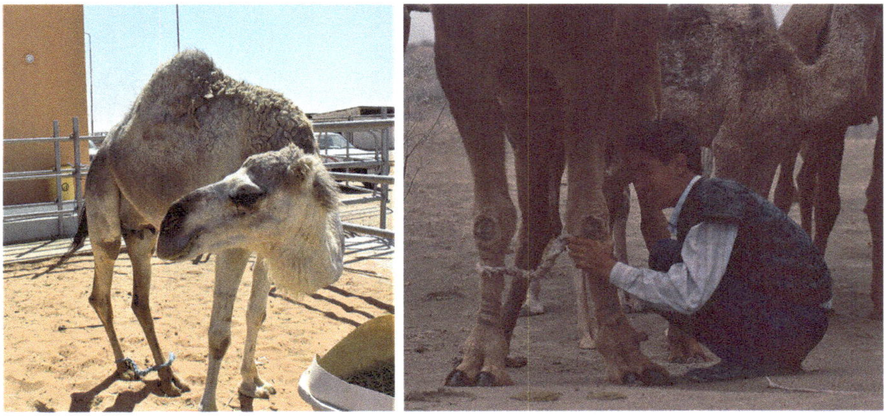

a b

Fig. 6.13 (**a**) Hindrance of two hind legs in a camel. (**b**) Hindrance of two forelegs in a dromedary

Fig. 6.14 Immobilization of a camel by forelimb restraint

Local anesthesia is used for light and localized interventions, incising an abscess, making a suture, and placing a catheter. It is often used as a complement to sedation, particularly in the case of castration or caesarean section. Except in young animals, overdosing is well tolerated with local (subcutaneous) injections. The most common

a b

Fig. 6.15 (**a**) Restraint cage for individual examinations. (**b**) Refusal of entry into a restraint cage by an untrained camel

Fig. 6.16 Holding corridor with electronic scale in Saudi Arabia

anesthetic is Xylocaine 2% (or similar forms). For anesthesia of the ocular region, tetracaine or proparacaine is used. For endoscopy or endotracheal intubation, lidocaine gels can be used.

The most delicate injection concerns epidural anesthesia, recommended in case of perianal surgery, but especially during reduction of a rectal or vaginal prolapse. It is performed in the first mobile intervertebral space after the sacrum (to locate it, one

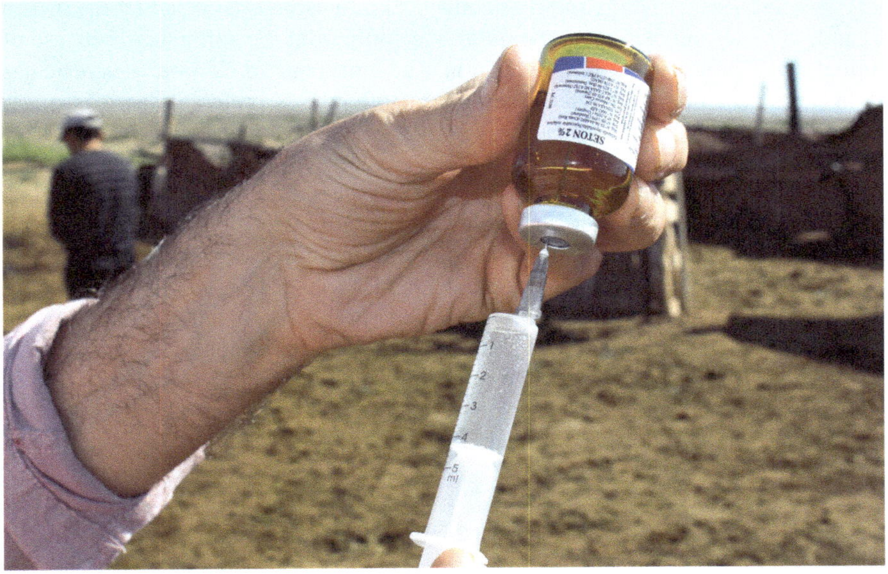

Fig. 6.17 Preparation of a sedative before surgery on a Bactrian camel in Kazakhstan. (Photo G. Konuspayeva)

must grasp the tail with one hand and pull the tail up to locate the mobile space). It is best to shave the area and disinfect it well to better locate the injection site. The needle is inserted perpendicularly or slightly forward. One to 3 ml of lidocaine HCl (without epinephrine) is injected for 1 to 2 hours of anesthesia.

General anesthesia requires more care. A preanesthetic blood test may be useful for elderly or sick animals. As the risk of regurgitation is high, it is useful to proceed with a solid diet for 24–36 hours accompanied by a hydrous diet for at least 12 hours. Anesthesia can be induced preferably in the lying animal after sedation, but it is also possible to do it in the tranquilized standing animal. If the animal lies on its side, it is preferable that it be on the right side, which is less likely to cause regurgitation. The head can be held slightly elevated to avoid passive regurgitation, but if regurgitation occurs, the head should be lowered as much as possible immediately to avoid false swallowing. Finally, to avoid the risk of paralysis, the forelimb below should be pulled to avoid crushing the radial nerve against the ribs.

There are two types of general anesthesia:

- Fixed anesthesia with a catheter placed in the vein through which it is injected, for example, a mix of ketamine 1 mg/ml associated with guaifenesin 5% with an induction at 1.7–2.2 ml/kg and then maintenance at 2.6 ml/kg/hour (+/– butorphanol 0.02 to 0.05 mg/kg) or a mix of ketamine 2–5 mg/kg and diazepam 0.2–0.5 mg/kg.

- Gas anesthesia, which requires endotracheal intubation after cleaning the mouth and, in males, repositioning the *dulla*, the soft part of the soft palate, because of the risk of protrusion during intubation. The most commonly used anesthetic gas in large camelids is halothane, although isoflurane is preferable as it is a less cardiorespiratory depressant.

To wake up, the animal should be placed in a sternal decubitus position with its head held straight. Straw bales can be placed to stabilize the animal. It is preferable to restrain the animal in this position and wait until it is fully awake to release it. The length and flexibility of the neck of camelids mean that an animal that is not fully awake can dislocate its cervical vertebrae when it wakes up by falling back badly on its neck. The head should be kept slightly elevated to limit regurgitation. Once the animal is well awake, a progressive refeeding can be quickly implemented (about 12 hours). If local anesthesia can be done by the breeder, in all cases general anesthesia must be followed by a veterinarian.

6.3 Clinical Examination and Specimens

It is recalled here that dromedaries and Bactrians are not very expressive of their pain and symptoms can be very frustrating. It is therefore essential to collect precise memoranda. The general data of the animal (age, living environment, presence of other animals, food, etc.), the behavioral data (general attitude, depression, apathy, aggressiveness, etc.), and especially the signs attesting to a pathological problem (cessation of rumination, lameness, diarrhea, hyperthermia, respiratory difficulties, nasal or genital discharge, etc.) should be recorded.

6.3.1 General Clinical Examination

A clinical examination consists of a careful look at several points:

- The animal's overweight status: there is a scoring grid for camel overweight status (Box 6.1). A very thin animal may indicate a parasitic disease (e.g., trypanosomosis), digestive problems, or poor feed management. Conversely, overweight animals may be affected by reproductive or locomotor disorders.
- The general posture: animal standing or lying down, in sternal or lateral decubitus, and agitated or apathetic behavior are all signs of fatigue or pain.
- The general appearance: normally, the hair is provided on the neck, the hump, and the flanks; it is very short, even absent in the ventral region and on the internal face of the thighs. A "pitted" or dull coat, the presence of edema or abscesses, traces of sweating, a moth-eaten aspect of the wool, the presence of irritations, and wounds are signs of "disorders." Note that during the molting period (shedding of the winter wool), the animal may look "shabby" without this being linked to any disease (Fig. 6.18).

Fig. 6.18 “Shabby” appearance of a hybrid at spring molt (natural shedding of winter wool)

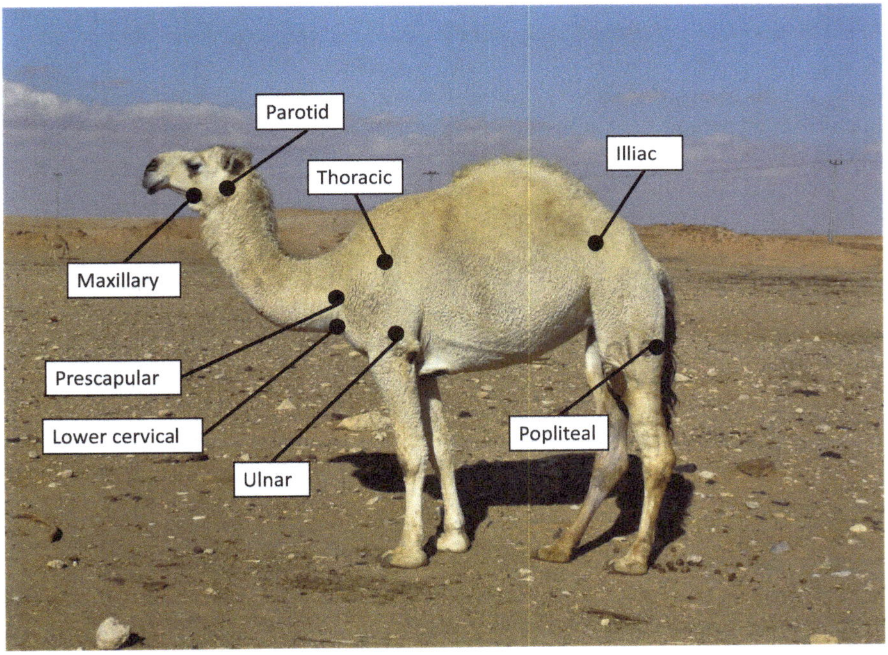

Fig. 6.19 Location of the main palpable nodes in the dromedary

- Examination of the lymph nodes: they are the first to become abscesses (Fig. 6.19).
- Eye examination: the eye should be wide open, without tearing, and its pupil is oval and horizontal with small *migrating* bodies on the dorsal and ventral margins. The iris may be discolored. The retina is without a mat or *fovea* with numerous vessels that converge toward the papilla. The cornea is smooth and shiny. Palpebral and pupillary reflexes are present. The lacrimal duct is 2–3 mm in diameter, extending from the medial canthus to the mucocutaneous junction of the nostril. The presence of tears, a closed eye, an opacity on the cornea, and an absence of reflexes are pathological signs. In dry weather, the eye is regularly moistened by the nictitating eyelid (the third lateral eyelid), whose function is, among other things, to lubricate the ocular surface and protect the conjunctiva from sandy winds.
- Examination of the nostrils: normally the nostrils are quite dry. Note the presence of uni- or bilateral discharge, serous or purulent, with or without odor, linked or not to a respiratory disorder.
- Examination of the oral mucosa: normally pink, it may also be irregularly black in a physiological way. Excessive pallor and congestion are pathological signs.
- Temperature measurement: beware, as large camelids have a temperature that varies during the day depending on the outside temperature, between 35 and 40 °C, without this being linked to an infectious problem. It is thus a parameter not very reliable.
- Heart rate: the stethoscope is placed behind the elbow, against the thorax, on the left. The rhythm is sinus with two distinct sounds and a frequency of between 30 and 50 beats/min. The arterial pulse can also be taken under the tail in young animals or behind the carpus in adults. An increase in frequency is observed in case of stress, but also in case of pain. A disturbance of the rhythm or a murmur is a sign of a heart disorder.
- Respiratory rate: the breathing of large camelids is normally silent and takes place through the nostrils. Thoracic movements are barely perceptible. To measure the respiratory rate, you must put your hand flat on the thorax to feel it or place the stethoscope on the trachea. A respiratory disease (pneumonia) is characterized by an increase in the amplitude of the thoracic movements, difficulties in ventilation, sometimes wheezing, and even coughing.
- Examination of the digestive system: by auscultating the abdomen on the left side, one perceives the noises at the level of the stomachs at a rate of 3–4 sounds emitted per minute. By tapping with the fingers also on the left side, a tympanic noise is perceived during gaseous distensions.

However, the main signs of disease in large camelids are droopiness and loss of appetite or even anorexia. In addition, the restraint sometimes necessary for certain stages of the clinical examination can distort the data collected (respiratory and cardiac rhythm, rectal temperature). It is therefore often essential to carry out additional examinations and therefore to take blood or other biological fluids (urine, milk) and, in the event of unexplained mortality, to use medical imaging techniques (ultrasound) and to carry out an autopsy.

Box 6.1 Body Condition Scoring Grid (Fig. 6.20)

6.3.2 Blood Sampling

Blood sampling from a standing animal should preferably be done with the neck stretched downward to facilitate venous stasis. The forelimbs are restrained. In more nervous animals, it is preferable to collect blood from the animal in a stance, as it is easier to collect blood from the neck bent against the thorax (see Fig. 6.9).

The sampling site on the jugular vein is easy to locate, especially after applying even slight pressure at the base of the neck or, preferably, halfway between the thorax and the head. The easiest sampling point is 15–20 cm from the mandible. However, in males, this anatomical region is endowed with abundant and long hair, which can make tactile perception of the jugular vein difficult. It is therefore necessary to shave or cut the excess hair beforehand. A confusion of the jugular vein with the distended part of the throat in the angry or stressed animal is not impossible. Care should be taken to feel the roll of the vein under the fingers before proceeding with the operation (Fig. 6.21a).

Blood can be drawn from other areas, such as the medial metacarpal vein (visible on the medial side of the carpus), the dorsal metatarsal vein (visible on the cranio-lateral border of the metatarsal between the extensor tendons), or the caudal vein, but these areas require severe restraints. In lactating females, however, it is easy to draw blood from the mammary vein, which is usually clearly visible (Fig. 6.21b).

The use of Vacutainer© tubes allows the use of thinner needles, less traumatic for the animal. Moreover, in dromedaries, the resistance of red blood cells is on average much higher than in other species, which considerably limits the risks of hemolysis. Consequently, it is not always necessary to use sophisticated blood sampling systems. However, the dromedary's reputation for rusticity should not obscure the need to follow the classic rules of hygiene when taking blood samples: disinfection of the skin with alcohol, possibly after cutting the excess hair on the intervention area, use of one needle per animal, or disinfection with alcohol between each use. In the latter case, it should be noted that the skin of dromedaries is very thick, sometimes indurated by frequent skin diseases (e.g., mange), and that the needles quickly dull.

The assays are performed either on whole blood, plasma, or serum. Depending on the parameters to be studied, an anticoagulant (heparin, liquemine, fluoride/oxalate, EDTA) is used or not. The plasma of large camelids is very translucent because of the low level of carotenoid pigments. If the analyses must be postponed, it is imperative to ensure that the samples are preserved in the best possible conditions. Particularly, enzymatic and hormonal parameters do not stand up well to a break in the cold chain. From a general point of view, there are no specific rules for blood sampling and storage of samples in camelids. The rules in question remain the same

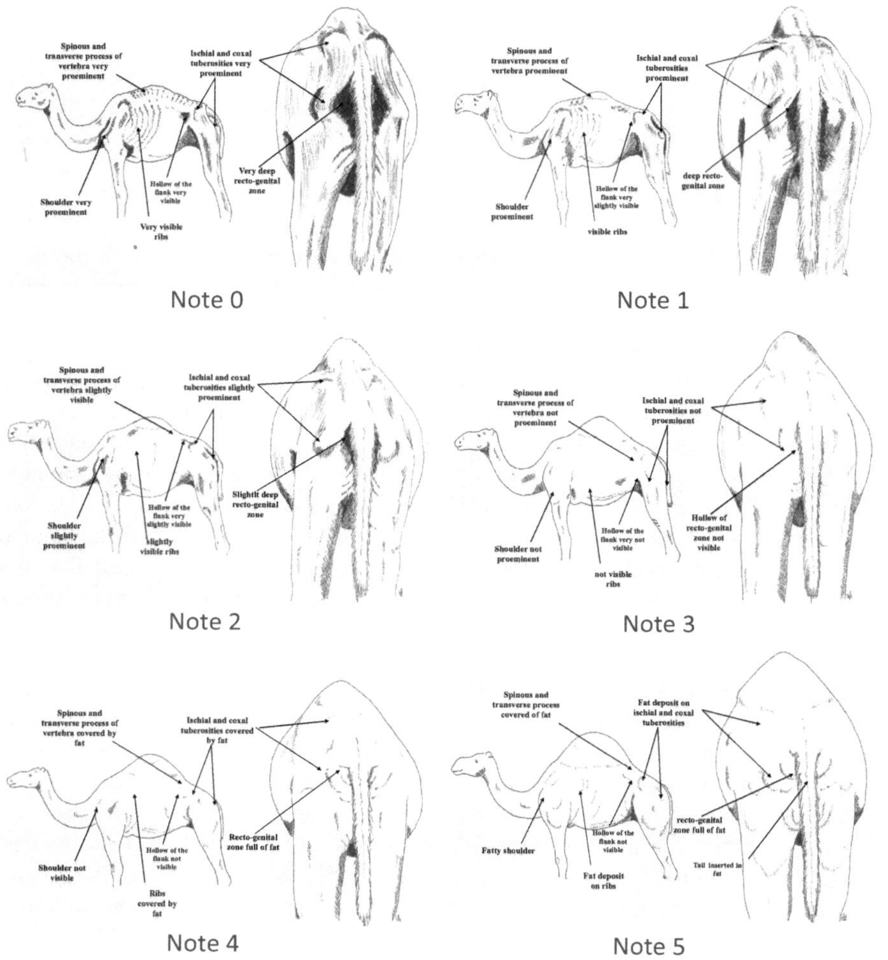

Fig. 6.20 Dromedary body condition scoring grid (score 0–5)

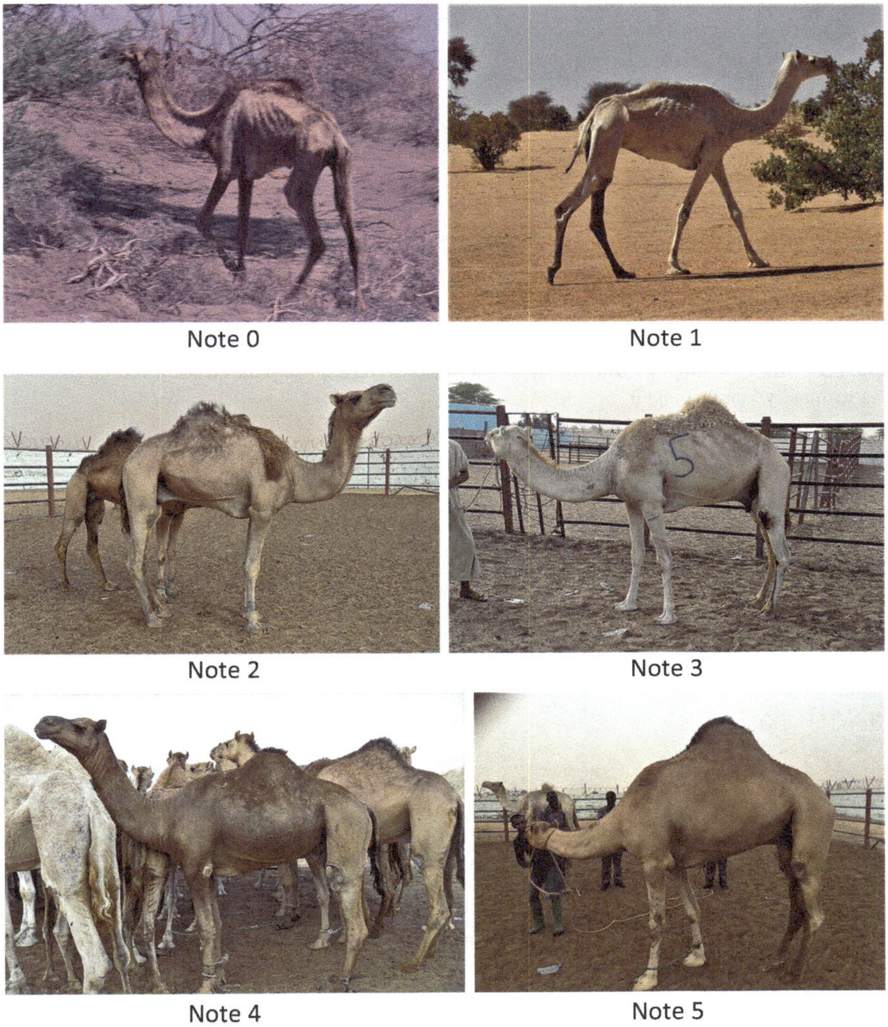

Fig. 6.20 (continued)

as those applied to other species of zootechnical interest. However, the ecology of the dromedary (arid and semiarid zones with sometimes difficult access) and its extensive breeding method (nomadism, transhumance) require additional precautions to ensure that samples are stored in the best possible conditions.

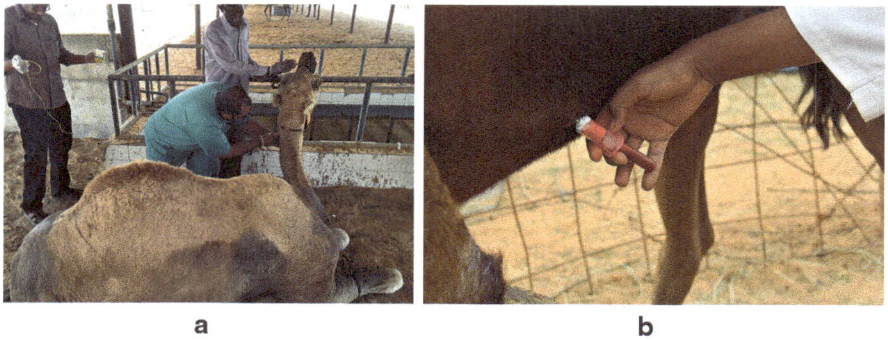

Fig. 6.21 Blood sampling. (a) From the jugular vein of the animal in the stance position. (b) From the mammary vein in camels

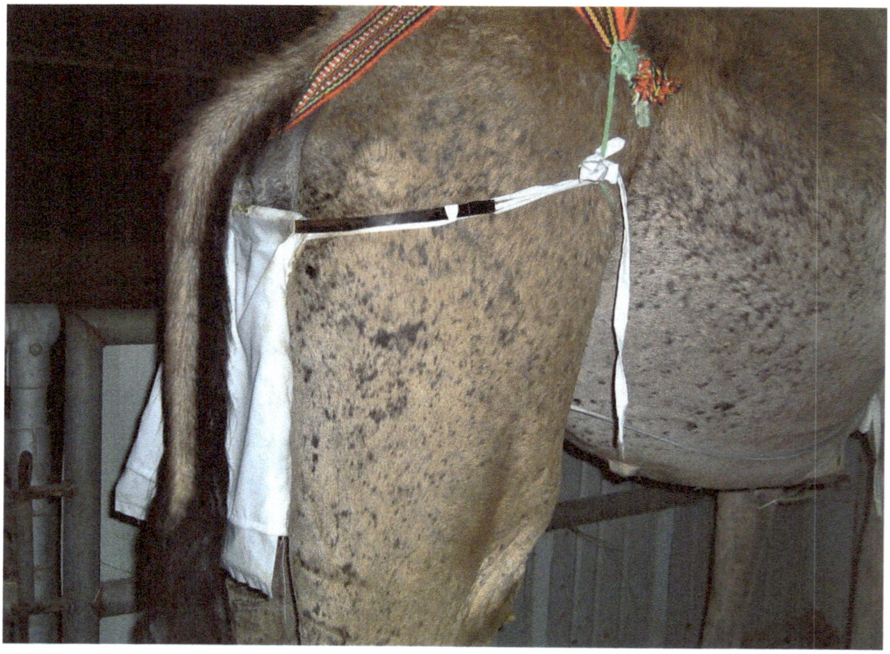

Fig. 6.22 Twenty-four-hour urine collection device in a camel

6.3.3 Other Types of Sampling

Collection of urine is less easy in animals than blood. The determination of biochemical parameters of clinical interest is therefore much less used, particularly in the dromedary, as the relevance of the results requires collection over a 24-h period. Urine can be collected by natural sampling or by using a urethral catheter, although

Fig. 6.23 Adult dromedary urinary stone. (Photo C. Magnan)

in females a diverticulum in the urethra makes catheterization difficult. For 24-h urine, there are collection techniques using a plastic bag adapted to the anatomy of the female (Fig. 6.22), but these are mainly of interest under experimental conditions.

Urine collection is used when there is a suspicion of kidney or bladder damage (bacteriology, enzyme analysis) or urolithiasis in males. The normal color of the urine is transparent or pale yellow depending on the degree of hydration. The pH varies between 6 and 9. The specific gravity is 1022–1070. Normal urine is free of blood and protein. However, the presence of calcium oxalate sediment (Fig. 6.23) or ammonium phosphate is normal. The urine of dehydrated camelids is highly concentrated.

Milk samples can be useful for bacteriological analysis, more rarely for biochemical assays (e.g., urea). Sampling is generally very easy in camels used for milking.

Fecal samples are taken for the evaluation of gastrointestinal parasitism (coproscopy) and possibly for bacteriological analysis. Fecal collection is easy. It can be performed directly in the rectum. Humidity is particularly low in dromedary excreta, so preservation is easy.

Other possible samples from internal organs (by biopsy), cerebrospinal fluid punctures, and recovery of rumen juice (by orogastric probing) are not widely used and require good technical skills. It is preferable to call on the veterinarian for these practices.

6.4 Interpretation of Analyses

In the context of the present work, we will limit ourselves to some broad outlines. For further details, reference can be made to books on clinical biochemistry in large camelids (Faye and Bengoumi, 2018). As in other livestock species, biochemical investigations, especially blood, are practical and relevant if (i) the purpose of the analyses is clear and based on an adequate sampling plan; (ii) physiological factors of variation are taken into account for a practical interpretation of the results; (iii) the sampling protocol is respected (e.g., fasting blood sampling); (iv) the interpretation is based on a set of animals because of variability between individuals and on a set of parameters (e.g., blood urea level should be interpreted in relation to the level of enzymes that indicate renal impairment and blood glucose); and (v) confounding between statistical difference and biological significance is avoided.

With regard to the specificities of large camelids, the following few elements can be recalled:

(i) The predominance of polynuclear neutrophils in the leukocyte formula
(ii) The maintenance of the hematocrit in case of physical effort
(iii) The resistance to osmolality
(iv) A relative hyperglycemia
(v) A quasi-absence of ketone bodies
(vi) Low plasma cholesterol level
(vii) Susceptibility to hyperuremia
(viii) Thermal resistance of alkaline phosphatases
(ix) Maintenance of metalloenzyme activities such as glutathione peroxidase or ceruloplasmin in case of mineral deficiency
(x) Maintenance of electrolyte balance in dehydrated animals
(xi) Low plasma zinc concentration in the supplemented camel
(xii) Greater sensitivity to selenium toxicity
(xiii) Vitamin C richness of milk
(xiv) Vitamin D richness of plasma
(xv) Seasonal pattern of testosterone
(xvi) Decreased insulin responsiveness with age of camels to induced hyperglycemia

A summary table of biochemical and hematological standards for large camelids is given below, taken from the abovementioned book (Table 6.1).

6.5 Diseases and Conditions

An exhaustive list of all the health problems encountered in large camelids can only be included in a book specifically devoted to this subject. As for biochemical and hematological analyses, we will refer to a few works available in the literature (Köhler-Rollefson et al., 2001; Wernery and Kaaden, 2002; Wernery et al., 2014), most of them in English, more rarely in French (Pacholek et al., 1999). We will therefore limit ourselves to the most frequent health problems giving rise to veterinary consultation.

Table 6.1 Usual values of camel biochemical and hematological parameters

Parameters	Usual values	Parameters	Usual values
Hematology		**Macrominerals and electrolytes**	
Red blood cells	$6–10 \times 10^6/mm^3$	Sodium	140–180 mmol/l
Hematocrit	25–30%	Potassium	3.5–6.3 mmol/l
Hemoglobin	9.3–15.5 g/dl	Chlorine	106–123 mmol/l
Hemoglobin	9.3–15.5 g/dl	Chlorine	106–123 mmol/l
MCV	30–45 fl	Bicarbonates	22–30 mmol/l
MCH	12–18 pg	Calcium	8.4–12.4 mg/100 ml
MCHC	40–50 g/dl	Phosphorus	3.8–8.4 mg/100 ml
White blood cells	$10.5–15.5 \times 10^3/mm^3$	Magnesium	1.8–2.8 mg/100 ml
Lymphocytes	29–63%	**Trace elements**	
Neutrophils	0–1%	Copper	70–120 µg/100 ml
Eosinophils	1.5–13.8%	Zinc	50–100 µg/100 ml
Monocytes	1–11.6%	Iron	70–120 µg/100 ml
Basophiles	<1%	Manganese	3–8 µg/100 ml
Plates	$230–360 \times 10^3/mm^3$	Selenium	50–150 ng/ml
Fibrinogen	200–400 mg/100 ml	Cobalt	30–60 µg/100 ml
Energy parameters		Iodine	50–120 ng/ml
Glucose	60–140 mg/100 ml	Fluorine	4–6 µg/100 ml
Ketone bodies	0.001–0.01 mmol/l	**Vitamins**	
Cholesterol	18–150 mg/100 ml	Vitamin A	20–50 µg/100 ml
Triglycerides	10–80 mg/100 ml	Vitamin B1	35–60 µg/l
Phospholipids	12–50 mg/100 ml	Vitamin B3	400–500 µg/100 ml
Nitrogen and protein parameters		Vitamin B7	20–50 ng/100 ml
Urea	8–30 mg/100 ml	Vitamin B9	0.5–1 µg/100 ml
Uric acid	0.2–2 mg/100 ml	Vitamin B12	20–30 ng/100 ml
Creatinine	0.8–2 mg/100 ml	Vitamin C	0.3–0.6 mg/100 ml
Total protein	6.3–8.3 g/100 ml	Vitamin D	70–90 ng/100 ml
Albumin	25–45 g/l	Vitamin E	50–400 µg/100 ml
Globulin	20–50 g/l	Vitamin K	20–60 ng/100 ml
Bilirubin	0.5–8.6 mg/l	**Hormones**	
Haptoglobin	0.1–0.6 g/l	Testosterone	2–35 ng/ml
Fibrinogen	2.2–3.6 g/l	Estradiol	9–110 pg/ml
Enzymes		Progesterone	1–9 ng/ml
ASAT	37–131 U/l	Prolactin	1–10 ng/ml
ALAT	6–25 U/l	Cortisol	3–30 ng/ml
ALP	32–110 U/l	Thyroxine (T4)	80–130 ng/ml
LDH	337–2620 U/l	Triiodothyronine (T3)	0.5–1 ng/ml
GGT	8–28 U/l	Melatonin	5–250 pg/ml
CK	40–120 U/l	Insulin	100–230 pg/ml
GLDH	0–97 U/l	Leptin	2.5–8 ng/ml
Ceruloplasmin	15–50 U/l	Calcitonin	70–180 pg/ml
GSH-Px	15–36 U/l	PTH	1.9–2.1 ng/ml
SOD	1400–1800 U/l	Osteocalcin	20–40 ng/ml
		Aldosterone	2–5 ng/ml
		Vasopressin	0.1–1 pg/ml

6.5.1 Skin Problems

Large camelids pay a heavy price for skin diseases, which are very common in this species. Mange, essentially sarcoptic (due to acarian called *Sarcoptes scabiei*), is probably the most widespread disease of camelids worldwide and relatively difficult to treat. When it occurs on a farm where it can affect 100% of the animals, it is difficult to get rid of because it is highly contagious and can even be transmitted to humans. Characteristics are scaly lesions on the head, legs, flank, and perianal area. In severe forms, more or less generalized hyperkeratosis (thickening of the skin) appears on the body surface (Fig. 6.24).

Diagnosis is made by microscopic identification after scraping the skin down to the blood dew (preparation with a drop of mineral oil under the slide). Observation × 40 is sufficient or by skin biopsy (8-mm punch biopsy at the lesions without cleaning).

Treatment involves improving farm hygiene where necessary, avoiding overcrowding and high density, keeping bedding clean, and disinfecting equipment. Animals should be checked before introduction to the farm. For curative treatment, topicals work well on chorioptic and psoroptic scabies, which are on the surface of the dermis, but are not very effective on sarcoptic scabies, which are deep in the dermis. Amitraz (250 mg in 1 l of water sprayed all over the body) or lindane (0.5% solution spray) and fipronil spray can be used. This external treatment should be combined with internal treatment with repeated injections of ivermectin and antibiotic therapy in case of superinfection (ivermectin: 0.3 mg/kg subcutaneous

Fig. 6.24 Sarcoptic mange in an adult dromedary

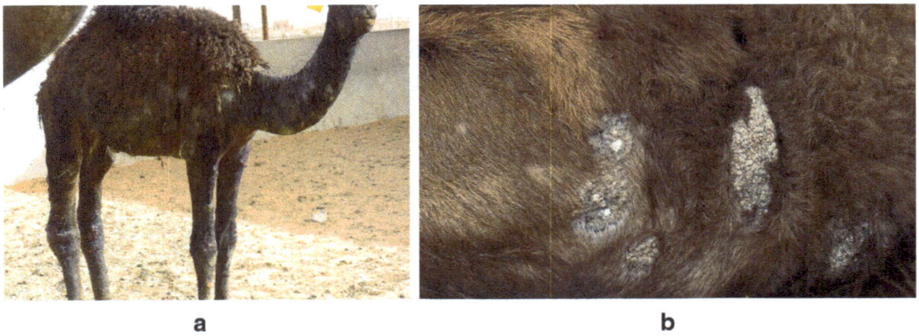

Fig. 6.25 (**a**) Young dromedary with *Trichophyton* ringworm. (**b**) Characteristic lesions of ringworm with "coin-like" depilation covered by a whitish crust

[SC] injection twice at 15 days' intervals; moxidectin: 0.5 mg/kg SC). Vitamin A and zinc supplementation is recommended. It is very difficult to treat in winter because of the abundant wool, and recurrences are common. Animals weakened by a pathology are more sensitive.

Less frequent, the genus *Chorioptes*, particularly *Chorioptes bovis*, is found generally on the foot and the inner side of the pasterns. Sometimes described on the head or the perineum, this parasite lives on the skin.

Ringworm is a fungal disease, i.e., caused by a fungus (*Trichophyton* sp.). It mainly affects young animals and, in most cases, disappears after 2–3 years. It is also highly contagious and causes characteristic lesions of "coin-like" depilations, covered with a whitish to grayish film, which are very pruritic and impair the growth of affected animals, which spend a lot of time scratching (Fig. 6.25). There are several forms of the disease, with varying degrees of severity.

Treatment is long and tedious, based on local applications of antifungal ointment (Imaveral®) and iodine products or by mouth (Dermogine®).

Abscess disease is also a very common pathology. It is caused by nonspecific pathogens (*Corynebacterium pyogenes* or *Y. pseudotuberculosis*, *Staphylococcus aureus*) often localized to lymph nodes, which swell and become painful. They may reach considerable volumes (Fig. 6.26a) and break through the skin, spilling copious amounts of pus.

Treatment is also local: after opening (debridement) and curettage, local disinfection (iodine) is performed. Large, well-demarcated abscesses can be surgically removed (Fig. 6.26b).

There are many other skin diseases, including viral diseases such as camelpox (due to camelpox virus), for which a vaccine exists, and contagious eczema and papillomatosis, for which the differential diagnosis is not always easy (Fig. 6.27) and for which treatments are not very effective, but these diseases are less common.

Ulcerative lymphangitis can affect all types of animals. This highly debilitating disease is caused by common germs such as *Corynebacterium* sp. or *Staphylococcus aureus*. It results in declining edema (in the limbs, then the genitals and abdomen)

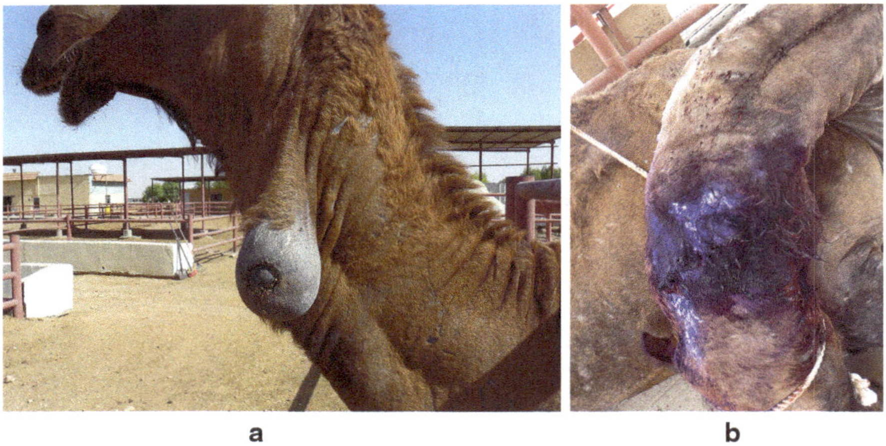

Fig. 6.26 (a) Skin abscess in an adult male. (b) The same animal after excision of the abscess

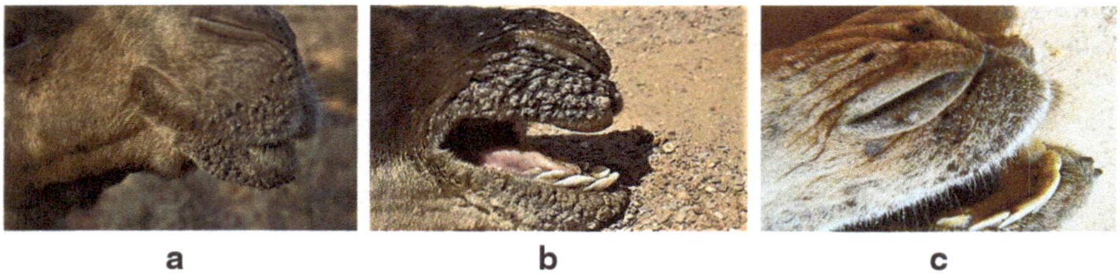

Fig. 6.27 Oral lesions of camelpox (a), contagious eczema (b), and papillomatosis (c). (Photos A. Thevenot)

followed by bloody ulcerative lesions (Fig. 6.28). Treatment includes antibiotics, anti-inflammatory drugs, vitamin A to regenerate the skin, and a diuretic (furosemide) to resolve the edema.

Cases of sternal pad necrosis have also been observed (Fig. 6.29), particularly in animals resting on traumatic, highly abrasive surfaces, causing microtrauma that subsequently becomes infected. After the abscess has been cured, a sort of sling should be made to position a cushion dressing. Indeed, the big camelids cannot absolutely remain upright, and the sternal position prevents the wound from healing; thus, it is necessary to protect the zone. A wheelbarrow tire attached to the front and back of the hump allows for support to be removed while leaving the wound open to the air.

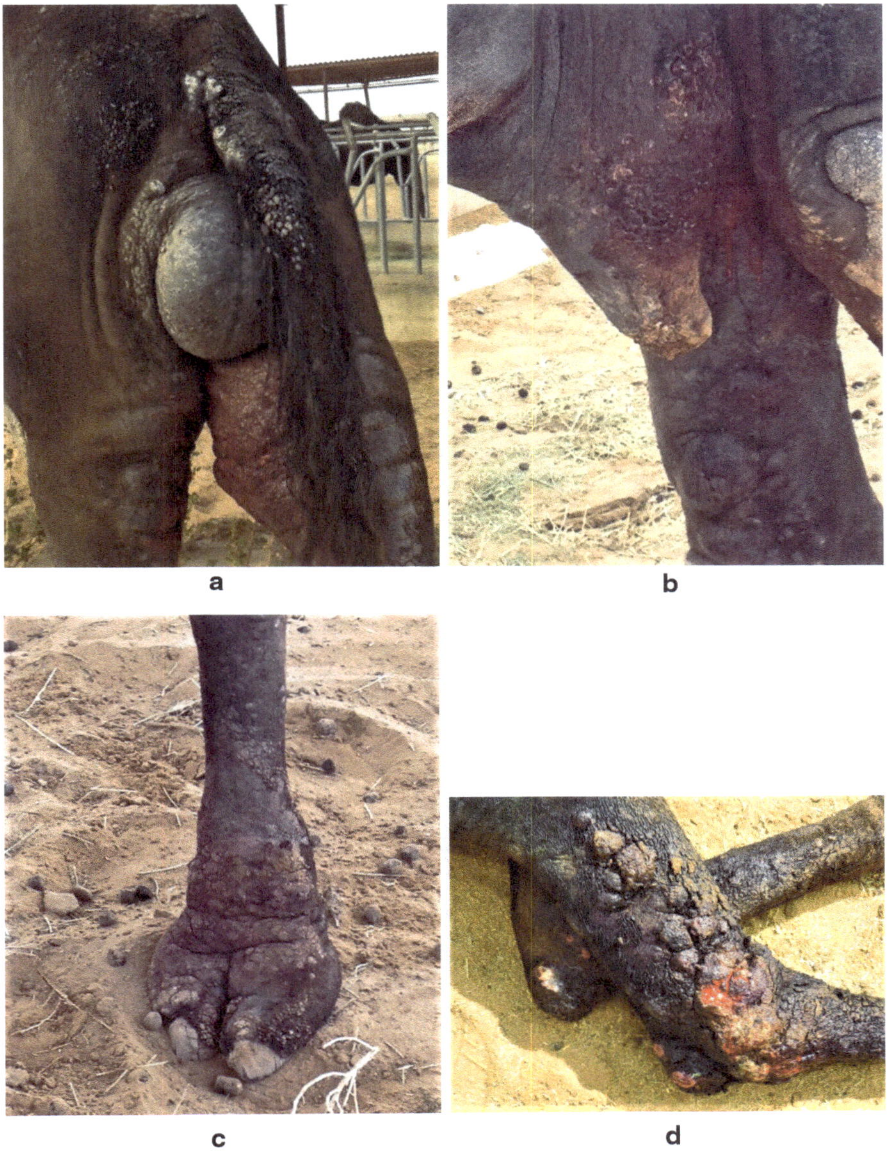

Fig. 6.28 Ulcerative lymphangitis lesions in an adult male. (**a**) Edema of the testes. (**b**) Edema of the penile sheath. (**c**) Edema of the distal limbs. (**d**) Bloody ulcerative lesions on the hind limbs

6.5.2 External Parasites

The main problem in most countries is tick infestation. There is one tick specific to dromedaries (*Hyalomma dromedarii*), but the animal may be parasitized by other species (Fig. 6.30a). Ticks are hematophagous acarian. They will therefore take

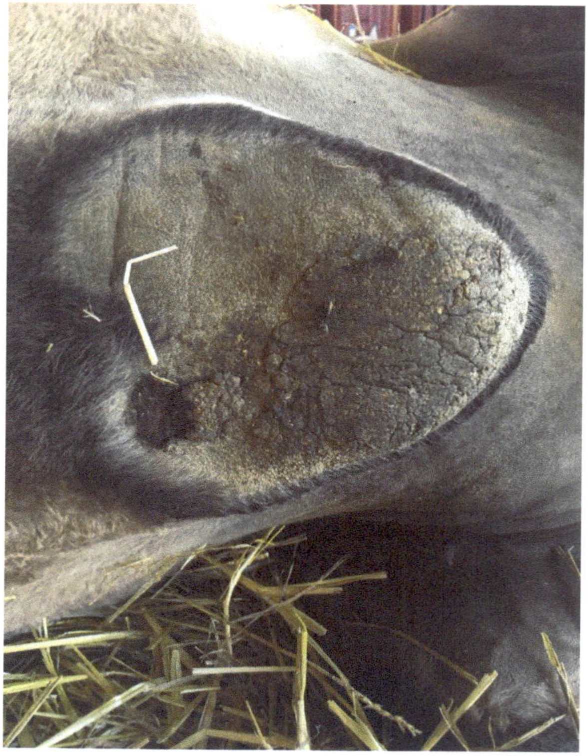

Fig. 6.29 Early necrosis of the sternal pad

a

b

Fig. 6.30 (**a**) *Hyalomma dromedarii*, the camel tick. (**b**) Ticks on the penis sheath of a male dromedary. (Photos N. Antoine-Moussiaux)

blood from the animal and can transmit or facilitate the onset of diseases such as those described above. Ticks are most often found in the most irrigated and thinnest areas of the skin (udder in females or sheath of the penis in males, nasal septum, anus, genital region) (Fig. 6.30b).

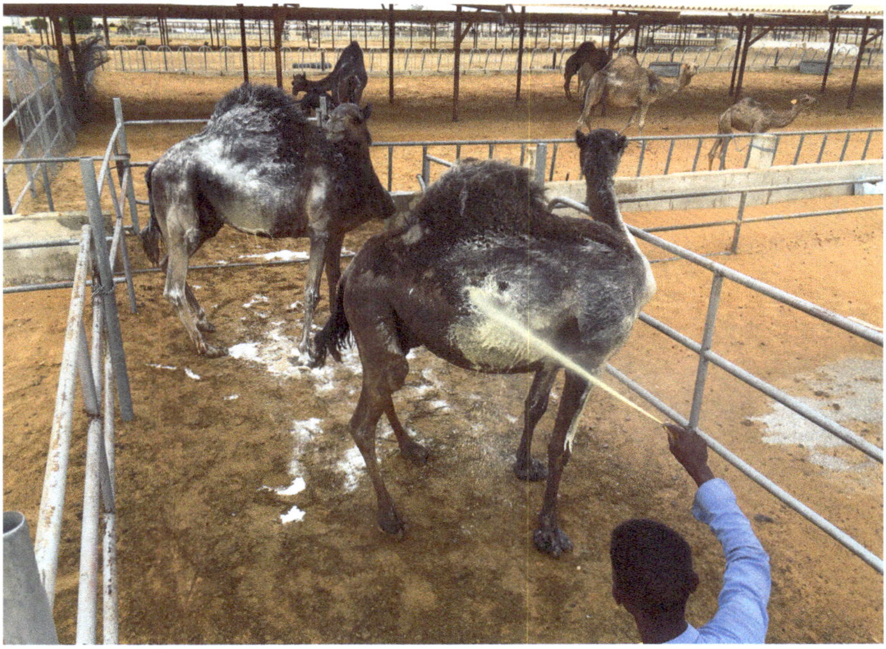

Fig. 6.31 Preventive treatment against ticks by spraying with a synthetic acaricide in Saudi Arabia

The treatment is based on acaricide spraying (Amitraz®) or pour-on (Ectosul®). Ivermectin (Ivomec®) also has a beneficial effect. The most effective preventive measure is to spray the animals regularly with a natural or synthetic acaricide (Fig. 6.31).

In certain climatic circumstances, certain anatomical zones of large camelids can be very sensitive to insect bites, causing skin reactions, significant itching, and conjunctivitis. A repellent treatment (local application of Dermistop®) is generally sufficient. A cortisone-based treatment can be applied in severe cases. In the Gulf countries, a mosquito net is placed on the heads of racing camels (Fig. 6.32).

The areas without wool are very sensitive to insects: head, limbs, and stomach. Bites can cause severe skin reactions, usually accompanied by severe itching. Tearing and in rare cases associated uveitis may also be observed (Fig. 6.33).

Lice are very difficult to see in the scales of the epidermis. They are about 4 mm in size and are white or brown. They are specific to camelids, with no intermediate host, and their life cycle is entirely on camelids. They are transmitted by close contact between individuals or by material (brushes, feces). Lice are of two types:

- Chewing lice. Preferential location: base of the tail and back. Strong itching; they cause a sparse coat.
- Biting lice. Preferential location: head, neck, flanks. Mild anemia possible.

Fig. 6.32 Mosquito net to protect the heads of racing camels from biting insects in Dubai

Fig. 6.33 Tearing in a camel due to biting insects

The treatment is based on ivermectin against biting lice (0.2 mg/kg in SC) and Sebacil®, Acadrex 60®, and Butox pour-on® against chewing lice, in external solutions.

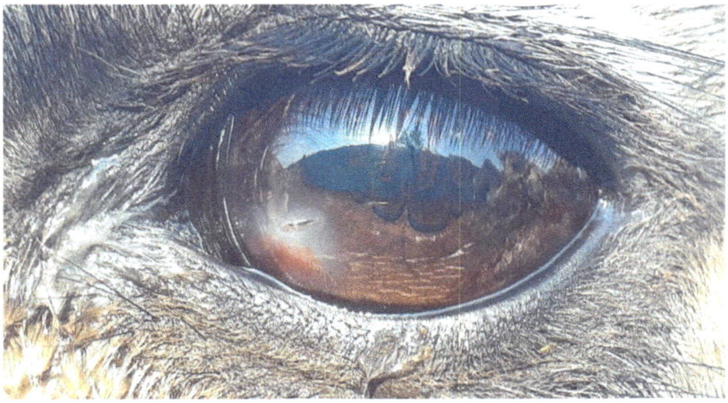

Fig. 6.34 Presence of a foreign body in the cornea. (Photo C. Lemeur)

6.5.3 Eye and Sinus Conditions

Ocular trauma, due to the height of the head in large camelids, is consequently less frequent than in other dairy species. In addition, the presence of the nictitating eyelid limits the risk of microtrauma due to sandstorms. However, following a shock or the presence of a foreign body, traumatic corneal ulcerations may occur (Fig. 6.34). Treatment is the same as for other species, i.e., removal of the foreign body or trimming of the ulcer, 1% atropine and local antibiotic ointment. In older individuals, especially Bactrians, it is not uncommon to observe spontaneous ulcers related to dry eyes caused by a failure to seal the occlusion line of the eyelids.

Uveitis has also been described, which can be treated with antibiotic eye drops but which may progress to cataracts. Congenital cataracts are not uncommon and may affect one or both eyes.

Conjunctivitis and keratoconjunctivitis of various causes are often a symptom of a general infectious disease. They may become purulent (Fig. 6.35). It should be noted that moxarellosis, which is very common in cattle and is transmitted by flies that stick to the eyes, has hardly been described in large camelids, whose tears are rich in lactoferrin, a protein with strong antibacterial properties. The treatment is based on eye drops and antibiotic gels. As the eye therapy requires intervention on the head, the animal must be well restrained. A subpalpebral catheter can be a practical solution for regular care of an inconvenient animal. A palpebral catheter used in equine medicine is perfectly suitable.

In some parts of the world, large camelids are frequently infested by the larvae of a fly (*Cephalopina titillator*) that parasitize the sinuses, where they cause irritation that is quite annoying to the animal (Fig. 6.36a). The main symptoms are sneezing with expulsion of the larvae (Fig. 6.36b) and breathing difficulties. In severe cases, these larvae can cause nervous disorders reminiscent of "whirling disease" in sheep with coenurosis.

Treatment with ivermectin (Ivomec®) is recommended. There are also traditional local treatments based on tobacco leaf.

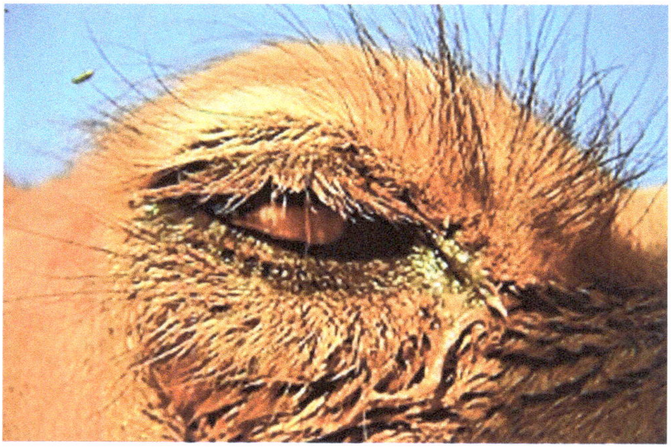

Fig. 6.35 Keratoconjunctivitis in a camel with purulent discharge. (Photo Cirad)

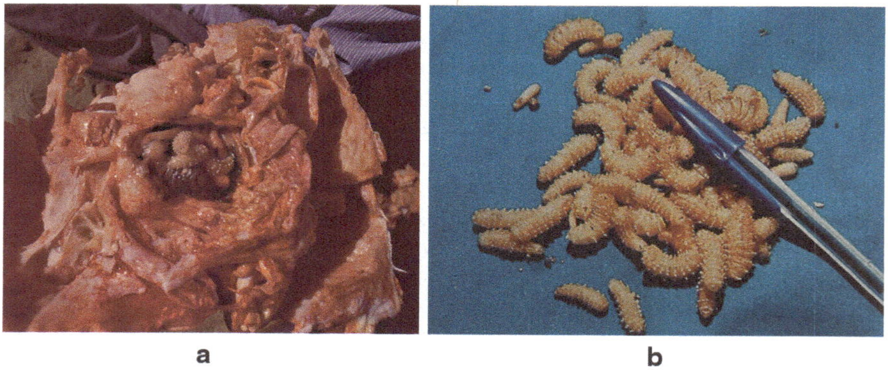

Fig. 6.36 Larvae of *Cephalopina titillator*. (**a**) In the sinuses of a camel in Turkmenistan. (**b**) Expelled after sneezing in adult dromedaries

6.5.4 Locomotion Disorders

Relatively less frequent in the countries of origin of camelids, locomotion disorders appear more frequent in European countries, where camelids have been imported and where their reduced numbers lead to inbreeding problems, such as defects in legs. With correct legs, the fetlock stands above the phalanges and does not touch the ground. With each step, it cushions the movement with the tendons that support it (Fig. 6.37a). In the case of hyperlaxity of the tendons, the fetlock collapses to the ground (Fig. 6.37b). This results in a misalignment of the phalanges; the nails are raised and no longer wear. The animal walks on its fetlocks. The result is great difficulty in locomotion with the animal suffering and unable to work.

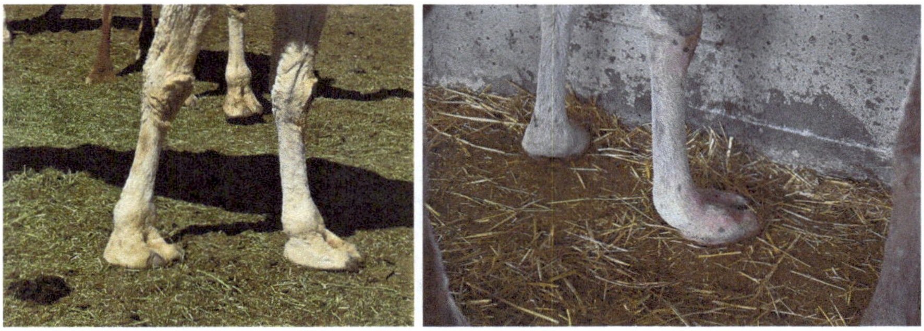

a b

Fig. 6.37 (**a**) Correct legs in a camel. (**b**) Incorrect legs in a young camel. (Photo C. Magnan)

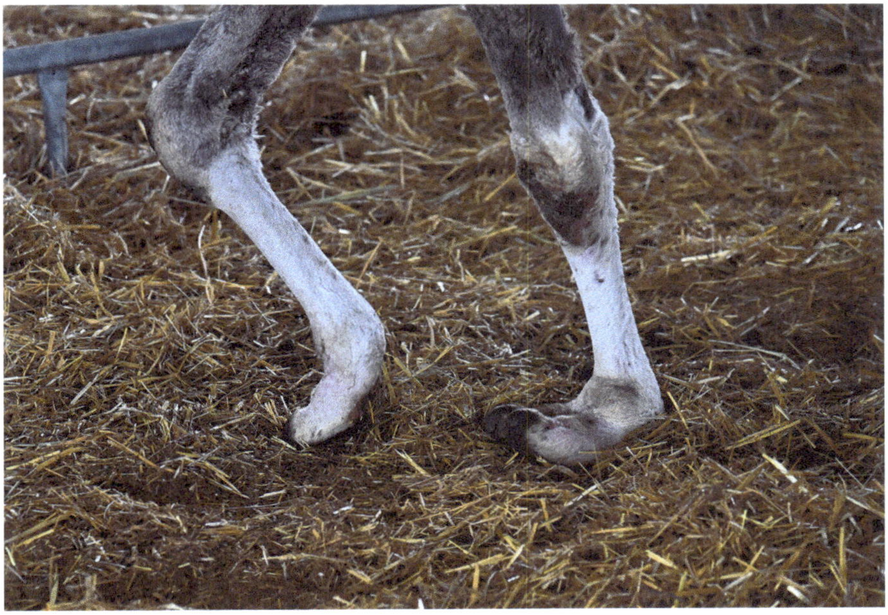

Fig. 6.38 Hoof deviation due to wear in a dromedary

In animals that do not move sufficiently (e.g., in indoor systems), the nails tend to grow uncontrollably, which can lead to deviated toes or to the nail becoming embedded in the sole (Fig. 6.38). It is then necessary to cut the nails that are too long to avoid lameness, sole abscesses, and even arthritis.

As the camelid foot has no hooves, trauma to the sole is not uncommon, especially on stony paths, even though the sole is protected by a thick, largely keratinized skin. In wetter countries where large camelids are newly established, the sole may be softer and therefore more fragile than in dry environments.

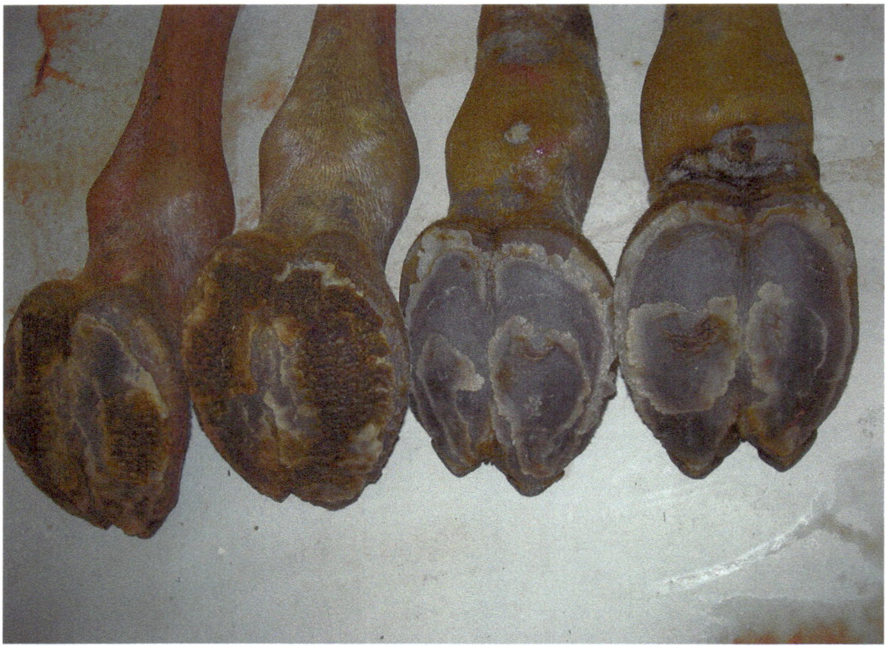

Fig. 6.39 Sole necrosis in a dromedary with selenosis in the United Arab Emirates. (Photo R. Seboussi)

Selenium deficiency or toxicity (selenosis) can also cause sole erosion with necrosis (Fig. 6.39).

In wet and cold soils, the feet of camelids can be strengthened with drying and tanning solutions (Solipat©, WinterPad ointment).

Fractures are uncommon in dairy farming. In males, however, jaw fractures are frequently reported after fights between animals in heat. For this reason, the cohabitation of several males in the same pen should be avoided. A surgical operation with a pin in the jaw is usually necessary (Purohit et al., 2019).

Rare cases of laminitis have been described in obese animals (high protein diet). One observes great difficulty in standing, a collapse of the phalanges. A diet and an anti-inflammatory treatment allow a return to normal if the case is not too severe.

Although not related to locomotion, the problem of idiopathic or traumatic torticollis can be recalled here. The anatomy of large camelids is characterized by a long neck that twists severely during unwanted contractions, causing severe pain and sometimes the inability to feed and drink (Fig. 6.40).

The cause is unknown and there is no possible prevention. Treatment includes an injection of an anti-inflammatory drug (dexamethasone) and a cocktail of B vitamins (e.g., Corebral®).

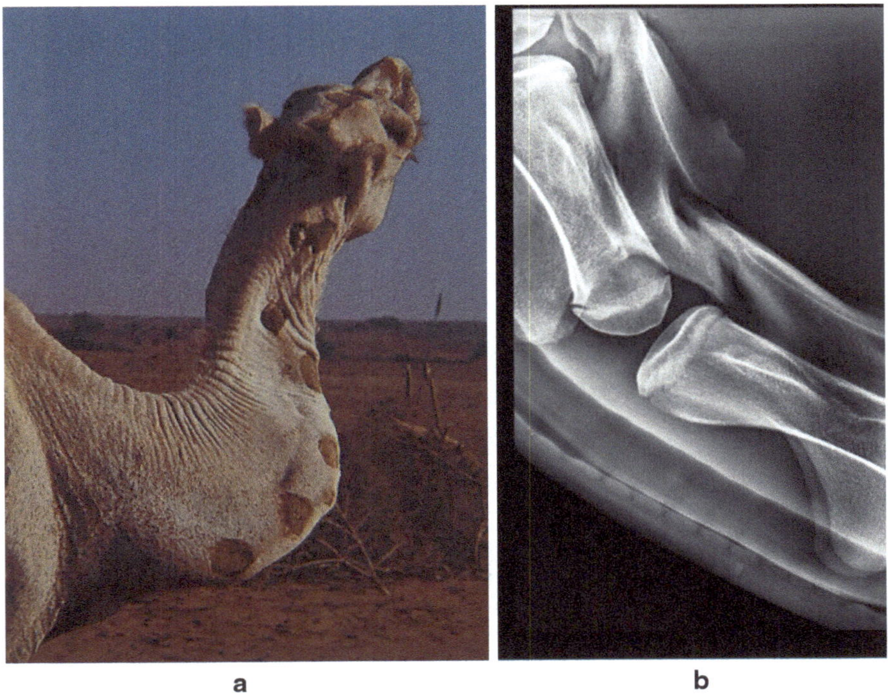

Fig. 6.40 (**a**) Torticollis in an adult dromedary. (Photo M. Dioli). (**b**) Traumatic dislocation of the cervical vertebrae in a dromedary. (Photo C. Magnan)

Table 6.2 Dosage of anti-inflammatory drugs used in camel medicine

Name	Dosage	Note
Anti-inflammatory drugs for acute treatment		
Dexamethasone	0.5 mg/kg IV	
DMSO	100 ml in 1 l NaCl IV	Better tolerated than in horses
Flunixin by meglunine	2.2 mg/kg IV	
Ketoprofen	2 mg/kg IV or IM	
Phenylbutazone	6.5 mg/kg IV slow or IM	4 packets of Equipalazone/day
Anti-inflammatory drugs for chronic treatment		
Harpagophytum	50–100 g per os	Generally associated with horsetail
Previcox	1 tablet (57 mg) per 500 kg per os	In cure or for life. First results after 8–10 days

Overall, camelids can be affected by lameness of various origins (arthritis, sprains, tendonitis, osteoarthritis) often treatable with anti-inflammatory drugs (dexamethasone, ketoprofen, or phenylbutazone) (Table 6.2).

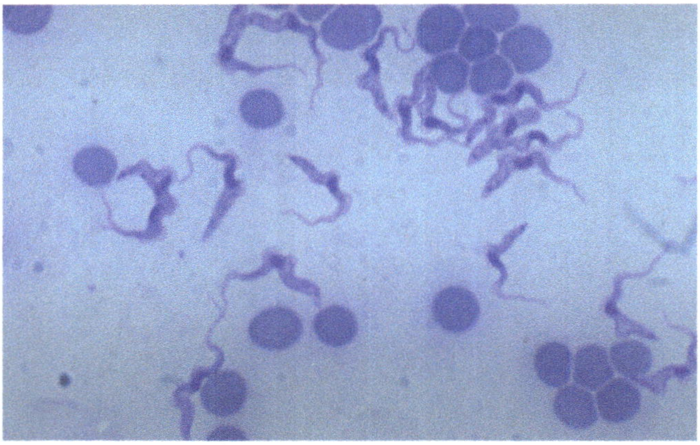

Fig. 6.41 Trypanosomes in the blood of an infected camel. (Photo CIRAD)

6.5.5 Trypanosomosis

Among the systemic disorders, internal parasitism appears very frequent in large camelids. There are two main types: blood parasitosis, especially trypanosomosis, and verminous parasitosis due to gastrointestinal worms (see section "Digestive Disorders").

Trypanosomiasis (*surra*) is one of the major plagues of camel farming worldwide. The disease is caused by a hemoparasite (*Trypanosoma evansi*) transmitted by biting flies such as stomoxes (Fig. 6.41).

The disease causes mainly depression, emaciation, and abortion in pregnant female. It is a very debilitating disease, the acute forms of which are fatal. The most efficient treatment is the injection of melarsomine (Cymelarsan®). It should be noted that, although this disease is not normally present in continental Europe, it may have been imported with animals from the Canary Islands (Desquesnes et al., 2007).

6.5.6 Digestive Disorders

Coccidiosis is often described in young animals, especially in stall farms. It is caused by a specific variety (*Eimeria cameli*), which causes severe diarrhea. However, many animals are asymptomatic. Treatment is based on anticoccidial drugs (e.g., SoluCox©).

The Bactrian camel (and especially small camelids) is also susceptible to cryptosporidiosis, which also causes diarrhea that can lead to death. The affected animal should be isolated quickly and treated symptomatically.

Another disorder is caused by digestive worms. Camelids are very sensitive to intestinal parasites, and, in their natural environment, the parasite load is very high. A parasitic infestation quickly leads to weight loss, diarrhea, and apathy which can lead to the death of the animal in case of massive infestation. The most common

Fig. 6.42 Roundworm *Haemonchus longistipes*, specific to camelids

Table 6.3 Dosage of the main antiparasitic molecules (again, word choice) used in large camelids

Molecules	Species	Doses	Targets
Albendazole	Bactrian	5–7.5 mg/kg PO	*Dicrocoelium*, nematodes
	Dromedary	7–10 mg/kg PO	No action on *Trichuris*
		20 mg/kg	Tapeworm, *Moniezia*
Fenbendazole	Bactrian	5–7 mg/kg PO for 1–3 days	Nematodes. Effective against larvae 50% *Trichuris*
	Dromedary	5–7.5 mg/kg PO	
Ivermectin	Bactrian	0.2 mg/kg PO 0.2 mg/kg HC	Nematodes ++strongyles *Trichuris*, ectoparasites
	Dromedary	0.2 mg/kg PO 0.2 mg/kg HC 0.5 mg/kg for-on	Nematodes Ectoparasites Caution: do not exceed 10 mg/kg in the young (lethal dose)
Levamisole	Bactrian	7.5 mg/kg PO	*Trichuris*, nematodes
	Dromedary	7.5 mg/kg PO	
Moxidectin	Bactrian	0.4 mg/kg PO	Nematodes
	Dromedary	0.2 mg/kg HC	Nematodes, *Sarcoptes*, ectoparasites
Praziquantel	Bactrian and dromedary	25 mg/kg PO	Tapeworm
Pyrantel	Bactrian and dromedary	25 mg/kg PO	Nematodes

PO per os – orally

parasite is haemonchosis, also caused by a camelid-specific variety (*Haemonchus longistipes*), a roundworm that lodges in the intestine (Fig. 6.42). Regular coproscopies are recommended for appropriate prophylaxis.

Flatworms such as tapeworms, *Moniezia*, echinococcosis, or *Dicrocoelium* also exist, but their frequency is lower, except locally. Many commercial antiparasitic molecules (not sure about the use of the word "molecules" here) can be used (Table 6.3). Rotation of molecules (same, again) during preventive treatments every 4 or 6 months helps to limit parasite resistance. Deworming a camelid is not always easy; that is why it is recommended to get them used to syringes filled with honey or applesauce in a calm way and from a young age.

Table 6.3 shows the most accessible molecules (word choice) in Europe with the dosage most often found in the various studies. A coproscopy allows a qualitative and quantitative assessment. However, it must be kept in mind that parasite excretion in feces is not regular. Therefore, false negatives can occur. In a herd, all the animals should be treated and not just one. Other species (small ruminants, cattle, horses, small camelids) that may cohabit with camelids should also be treated regularly.

One of the main health problems in camel farming is diarrhea in young camels, the causes of which are variable (bacterial, viral, parasitic, toxic). It is the main cause of camel mortality. In adults, diarrhea may also be associated with feeding errors or severe stress. Several specific diseases can result in enteritis of bacterial (enterotoxemia, salmonellosis, paratuberculosis, colibacillosis) or viral (rotavirus, coronavirus) origin. It is important to identify the origin of the diarrhea, to know the ration, to know the vaccination protocol (clostridium), and to know the deworming of the animal. A macroscopic and bacteriological analysis of the feces is very useful (color, smell, presence of mucus or blood). Treatments combine the following:

- Rehydrating agents (especially in young camels)
- Anti-acidosis products: sodium bicarbonate in 5% solution at a rate of 5 l/adult IV, effective in the early stages of endotoxemia
- 1.3% solution in NaCl with dextrose up to 60 l/24 h or 500 g p.o.
- Broad spectrum antibiotics
- Antispasmodics

The frequent reason for consultation, diarrhea is the sign of a disturbance of the intestinal flora and of an inflammation which can quickly lead to the death of the animal. Their origin can be diverse: parasitism, infections, and feeding errors. Feeding errors are generally a too high intake of nitrogen or proteins. Always check the animal's diet when diarrhea occurs.

Constipation is less common, but verminous obstructions are possible. In camels kept in cohorts in too small enclosures, intestinal obstructions by hair aggregates (pilobezoards) due to untimely licking of conspecifics can occur. Constipation can be recognized by symptoms of colic, lack of fecal excretion, and abdominal swelling. Treatment with antispasmodic products (e.g., Flunixin©) and paraffin oil can be recommended.

Some poisoning, especially from plants, can also cause digestive problems. These cases are to be feared when the animals change environment and are fed at pasture. The most frequent intoxications are linked, in Mediterranean environments, to the consumption of oleander, rhododendron, common sorghum, and boxwood. For example, the ingestion of rhododendron, an ornamental plant found in gardens, causes signs that appear within an hour: the animal starts to drool, lies down, hits its stomach, sometimes regurgitates, refuses to eat, and remains prostrate. Antispasmodics (Spamizole®, Calmagine®) should be administered intramuscularly or intravenously. An infusion of ringer/lactate helps to support the renal and

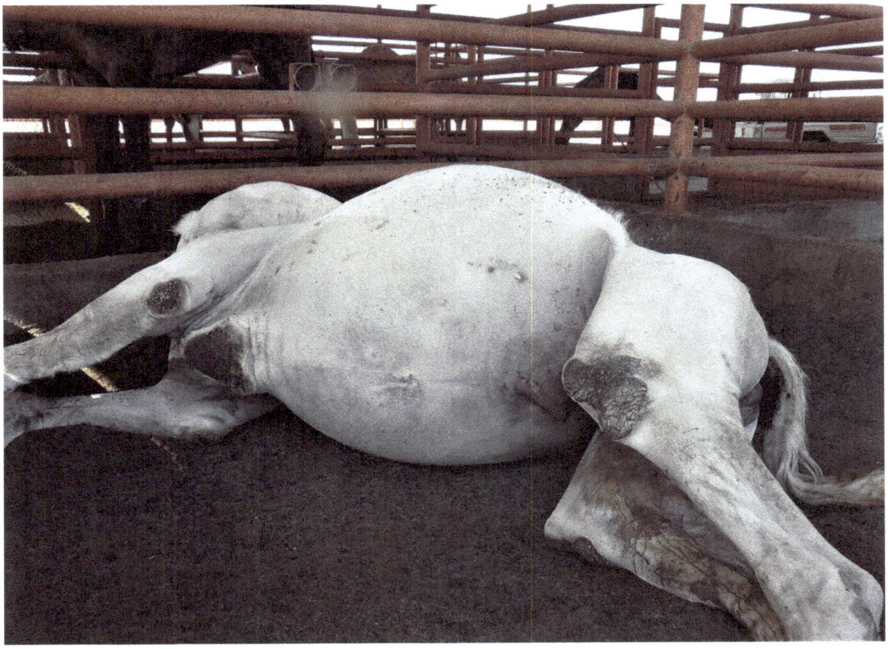

Fig. 6.43 Case of acute meteoritis in a 4-year-old dromedary fed a clover-based ration

hepatic functions. Recovery takes a few days. Ingestion of common sorghum (*Sorghum bicolor*), a plant grown for grain or fodder, is loaded with cyanide during its growth (like millet). Once cut and dried, this component disappears, and the plant can be ingested. If animals eat it fresh, it is toxic. They quickly develop severe colic with abdominal swelling, respiratory distress, and death. Treatment is symptomatic. The animal is probed to rinse the stomach with C1, and then paraffin oil and activated charcoal with 25 g of ammonium carbonate are added. If the meteoritis is very severe, the rumen can be trocaded. However, this procedure is much less well tolerated than in cattle.

Other cases of meteoritis have also been described. Due to the accumulation of gas in the stomachs because of feeding too much certain types of fodder (e.g., clover), this digestive disorder requires urgent intervention by trocarization of the stomach wall (Fig. 6.43).

6.5.7 Respiratory Problems

Like digestive disorders, respiratory disorders are most often of multifactorial origin. On normal clinical examination, respiratory movement is very discreet in large camelids, and a small serous discharge is frequently observed. On auscultation, air movements are not audible. In a sick animal, the most frequently observed

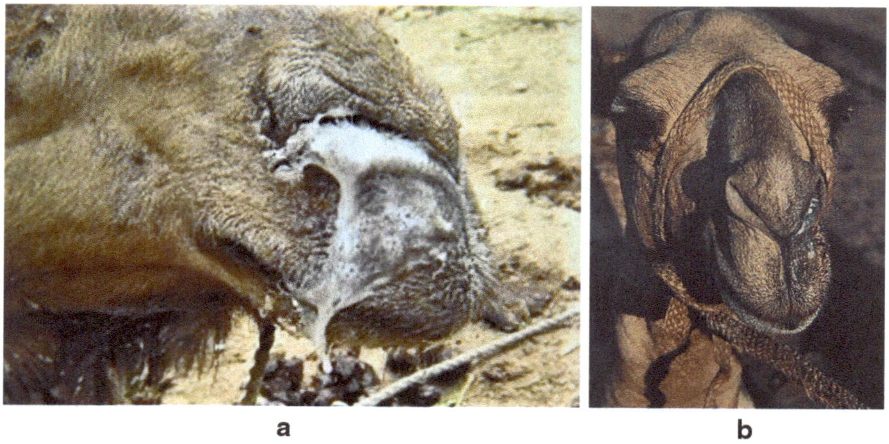

Fig. 6.44 Nasal mucopurulent discharge in camels with respiratory disorders. (Photo CIRAD and M. Dioli)

symptoms are coughing, nasal discharge (Fig. 6.44), and chest indrawing, which is more or less marked depending on the intensity of the respiratory distress. On auscultation, bronchial murmurs and wheezes may be heard. The differential diagnosis is not always easy: emphysema, infectious pneumonia or bronchopneumonia, pulmonary edema, pleurisy, or hydatid cysts.

Additional examinations are often necessary to adapt a treatment. A nasal swab during discharge can isolate a pathogen, and a pulmonary ultrasound will show pleurisy, abscesses. For complex cases, a transtracheal lavage will identify emphysema or an infectious or fungal pathogen.

Transtracheal lavage is performed under sedation and local anesthesia. On a calm and well-trained animal, a local anesthetic is sufficient. A small incision is made between two tracheal rings in the upper region, and a sterile sampling cannula is introduced into the trachea. A rinse with 40 ml of lidocaine is performed before collecting the rinsing liquid at the level of the low inflection of the neck. This simple technique allows the cause of the disease to be targeted and an appropriate treatment, antibiotic or otherwise, to be applied.

Specific diseases causing respiratory distress include pasteurellosis, but in most cases the causes are related to common environmental germs.

6.5.8 Reproductive Pathologies

See Chap. 5, section "Assisting Parturition," for aspects of dystocia.

One of the major postpartum problems is uterine prolapse. This is the externalization of the uterus (Fig. 6.45). Its occurrence requires rapid intervention, which consists of cleaning the organ and bringing it back in and then, after reduction, putting antibiotic tablets and suturing the vulva for about 10 days. Sedation, or even

Fig. 6.45 Post-parturition uterine prolapse in a camel

an epidural anesthetic, is recommended. To facilitate the reinsertion of the matrix, it is preferable to pull the hind legs backward (in the frog position) in order to induce a tilt of the pelvis. If the prolapse is accompanied by a rupture of the uterine wall with externalization of the intestinal loops, this leads to death in almost all cases.

There are many other pathologies both in females (vaginitis, metritis, pyometra, ovarian cysts, etc.) and in males (phimosis, orchitis, balanoposthitis, testicular trauma, etc.).

The most common problem in males is urolithiasis, which is the presence of urinary stones. The disease is manifested by signs of abdominal discomfort, anorexia, and depression. In advanced stages, this can lead to bladder rupture, which usually results in death. Treatment involves catheterization of the urethra (difficult because of the sigmoid flexure of the penis) or surgery. A urethrostomy is performed just opposite the ischium bend to place a catheter directly into the bladder. Ideally, the urethrostomy should be temporary, the time to remove the obstruction. In cases where this is not possible, it may become permanent with a risk of stricture of the urethra post-surgical. Prevention is based on a balanced diet and abundant watering.

Mastitis (see Chap. 5, section "Milking Hygiene") is the main problem in dairy farming (Fig. 6.46). The causes are multifactorial. A distinction is made between clinical mastitis (visible by symptoms of redness (tumor), warmth (calor), and pain (dolor) in the udder) and subclinical mastitis, which can be detected with, e.g., the CMT test (see Box 5.1). The latter are usually associated with minor pathogens. Severe forms (e.g., colibacillosis, mastitis) are accompanied by general signs (fever, depression, production fall) and a change in the milk (lumps, blood, even pus).

Super acute forms such as gangrenous mastitis can lead to death. Treatment is either local (injection of antibiotic into the teat canal, as the teat canal is not unique,

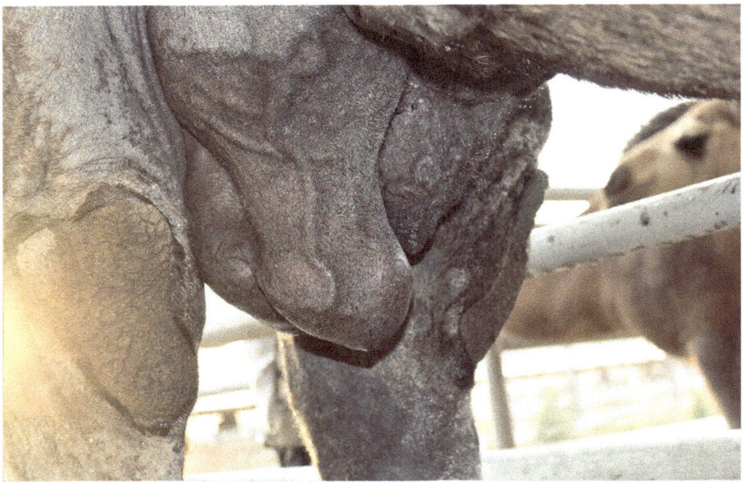

Fig. 6.46 Case of clinical mastitis in a quarter of a dairy camel, marked by a swollen teat

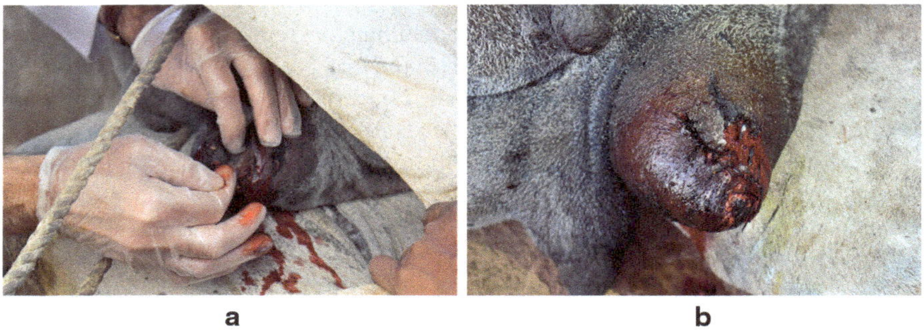

Fig. 6.47 Tear of the teat canal in a lactating camel. (**a**) Suture of the teat canal. (**b**) Finished repair. (Photo G. Konuspayeva)

unlike in cattle) or general (IV or IM antibiotic). After parturition, mammary edema is not uncommon. They resolve after administration of a diuretic and anti-inflammatory treatment. However, the frequency of mastitis is lower in dairy camels than in dairy cows.

Trauma to the udder is not uncommon either. They can lead to deep tears in the teat canal. Only surgical intervention is recommended (Fig. 6.47). Incorrect use of the milking machine (e.g., overmilking) can also lead to lesions (eversion), which can be an entry point for pathogens that cause mastitis.

Abortions also have multiple causes: infection, trauma, stress, intoxication, and parasitism. They can occur at any time during gestation (Fig. 6.48). If an infectious cause is suspected (brucellosis, Rift Valley fever, salmonellosis, Q fever,

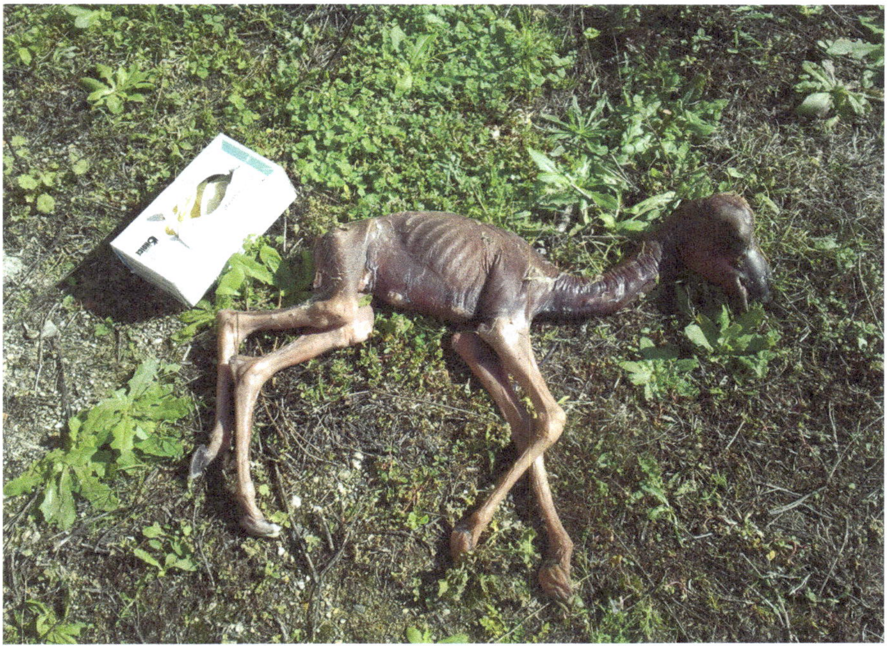

Fig. 6.48 Dromedary aborted fetus, about 6 months' old. (Photo O. Philipponneau)

chlamydia), it is important to make a serological diagnosis and possibly vaccinate the herd.

6.5.9 Major Infectious Diseases

Large camelids can be affected by most infectious animal diseases, whether bacterial (anthrax, botulism, leptospirosis, paratuberculosis, brucellosis, tuberculosis, tetanus, listeriosis), viral (rabies, smallpox, bluetongue, Rift Valley fever), or parasitic (toxoplasmosis, strongylosis, onchocerciasis), to name but a few. For further information, please refer to the books mentioned above. The Office International des Epizooties (OIE) in Paris has established through its ad hoc group on camelid diseases an exhaustive list of these diseases, diagnostic techniques, and therapeutic or preventive methods to be implemented, which allowed the publication of the book by Wernery et al., 2014. In all cases, the diagnosis of these diseases requires the intervention of vet practitioners and state veterinary services, as some of these diseases have important public health implications as zoonoses.

A final point concerning health management on a dairy camel farm is the fact that most of the diseases listed above are related to a health situation that is mainly encountered on farms in the South. The development of camel farming in Europe poses a real challenge for veterinary practitioners and managers, because camel diseases in a temperate climate, a particular social context (low mobility of herds in particular), and above all a completely different dietary context may take forms that have been little studied, or even never observed, particularly in terms of metabolic disorders.

6.6 Necropsy

Necropsy of camels that have died of disease is rarely performed in practice. The main reason for this is probably the size of the animal, which requires a suitable necropsy kit, and the lack of experience of most veterinarians and breeders in this field. However, a proper autopsy could provide valuable information on the cause of death of the animal when no diagnosis could be made, especially since, as mentioned above, the symptomatology is often frustrated. It is not uncommon to see a dromedary or a Bactrian depressed or anorexic and then to find it dead a few hours later without any characteristic sign being detected.

For complete information on necropsy technique and postmortem examination procedures, see Stimmelmayr et al. (1992) in the book of Schwartz and Dioli (1992). Autopsy involves the removal of all organs after opening the abdomen and thorax and possibly other anatomical regions (e.g., the skull and mouth) and inspection of these organs for abnormalities that may assist in diagnosis. The visual necropsy analysis can be completed by taking samples for laboratory analysis (microbiology, histology in particular). For more information on organ anatomy and the normal histology of the main tissues and organs, please refer to Jarrar and Faye (2012).

The standard procedure for a large camelid necropsy is as follows:

- Place the dead animal on the left side
- Incise the skin ventrally from the sternal pad to the udder or the penis sheath depending on the sex (Fig. 6.49a).
- Incise the skin and connective tissue links between the thorax and the shoulder in order to free the thorax.

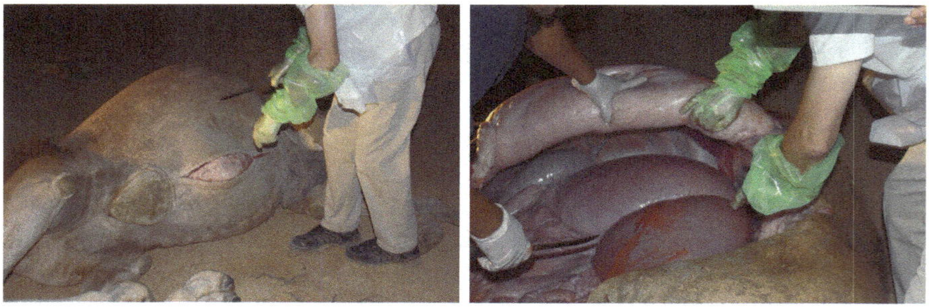

a b

Fig. 6.49 (**a**) Ventral incision of the abdomen during a camel autopsy. (**b**) Incision of the abdomen to remove abdominal organs. (Photos G. Konuspayeva)

Table 6.4 List of parameters and criteria to be recorded during a full autopsy

Element	Parameter	Criteria
Whole animal	Fattening condition	Obese, skinny, cachectic
Skeleton	Bones	Abnormalities, fractures
Skin and membranes	Color	Color anomalies
	Condition of the skin	Presence of hair loss
Muscles and fat	Color	Discoloration
	Consistency	Skinny, decaying
Internal bodies	Color	Discoloration, congestion (Fig. 6.50a, b)
	Weight	Density
	Form	Deformation
	Consistency	Indurated, necrotic, deliquescent (Fig. 6.50c)
	Topography	Arrangement of the organs
	Connective links	Adhesions
	Foreign bodies	Presence of a foreign body
	Smell	Abnormal, unpleasant
	Fluid in the abdomen	Presence of liquid (Fig. 6.50d)
	Lesions	Abscess, ulcer, tumor, hemorrhage, cysts
Structural change	Depth	Superficial/deep
	Distribution	Localized/generalized
	Size	Low/abundant
	Form	Firm/soft/liquid
	Quantity	Abundance of lesions
	Consistency, texture	Smooth/rough
	Odor and color	Abnormal
Parasites	Type	Ticks, worms, hydatid cysts
	Abundance	Numerous/invasive
	Location	Erratic location
	Parasite status	Living/dead

- Incise between the lower jaw and the base of the neck to access the cervical area.
- Incise the abdomen, taking care not to cut the intestines and stomachs (Fig. 6.49b).
- Cut the ribs near the vertebrae and the sternum and lift the rib flap.
- Eviscerate the entire lung-heart complex.
- Eviscerate the entire gastrointestinal tract, liver, and spleen.
- Remove the kidneys, genital tract, and adrenal glands
- Incise the joints and open them.
- Disarticulate the atlanto-occipital joint, open the skull, and extract the brain.

The following table (Table 6.4) lists the items that should be examined during an autopsy.

Every necropsy must result in a report that includes at least the following points:

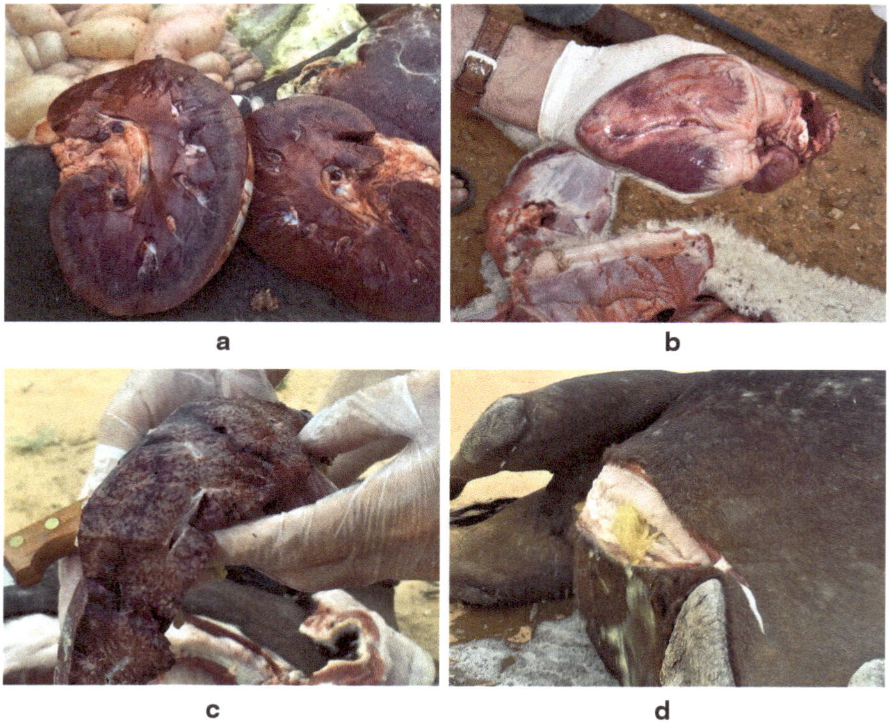

Fig. 6.50 (**a**) Kidney congestion. (**b**) Discoloration of the heart (selenium deficiency). (**c**) Liver necrosis. (**d**) Abundant fluid at the abdominal opening (ascites)

- Date and place of death
- Type of animal (breed, age, sex)
- General conditions (e.g., presence of visible external lesions)
- Commemoratives (symptoms, treatment received)
- Inspection of internal organs, lymph nodes, skeleton, muscles, brain
- Diagnosis
- Recommendations if necessary

References

Desquesnes M, Patout O, Brugidou R, Faye B, Cuny G (2007) Un foyer de *Trypanosoma evansi* observé pour la première fois en France, Bull. des GTV, 39, 8–10

Faye B, Bengoumi M (2018) Camel clinical biochemistry and hematology. Springer, New York, p 346

Jarrar B, Faye B (2012) Normal pattern of camel histology. FAO, Riyadh, Camel Breeding, Protection and Improvement Center, project UTF/SAU/021/SAU, 140 p

Köhler-Rollefson I, Mundy P, Mathias E (2001) A field manual of camel diseases. SCTD Practical Action Publishing, Bourton, p 254
Pacholek X, Vias G, Faye B (1999) Traitement des maladies du dromadaire. Guide de l'auxiliaire d'élevage. Niamey, Karkara publ. http://camelides.cirad.fr/fr/publications/guide_maladies.html
Purohit S, Kumar V, Kumar G, Pandey RP (2019) Management of mandible fractures using interdental wiring (IDW) in camels. J Camel Pract Res 26(2):189–190
Schwartz H, Dioli M (1992) The one-humped camel in Eastern-Africa. Verlag, Weikersheim, p 282
Stimmelmayr R, Dioli M, Schwarz HJ (1992) Field guide to post-mortem examination (chap. 5). In: Schwartz H, Dioli M (eds) The one-humped camel (C. dromedarius) in Eastern Africa. Verlag, Berlin, pp 225–262
Wernery U, Kaaden O-R (2002) Infectious diseases in camelids. Blackwell Sciences, Berlin, p 404
Wernery U, Kinne J, Schuster RK (2014) Camelid infectious disorders. OIE, Paris, p 500

Economic Management of a Dairy Camel Farm

7

Abstract

The question of the profitability of camel farm is currently asked. For assessing the ability of camel breeders to get sufficient incomes from their camel, different items must be taken into account. The central element for evaluating camel farm economy is the numerical productivity. In a farm with identified animals (different identification systems can be applied) and where the performances are recorded in one specific database, it is possible to foresee the demographic growth of the camel herd for subsequent years by knowing the main zootechnical items (fertility rate, mortality rate, culling rate, etc.). The evaluation of the economic performance of the farm includes the assessment of products and charges. The charges are composed of fixed expenses (amortizing rate of infrastructure and equipment, taxes and bank fees, administrative costs including staff) and variable charges (manpower linked to animal management, feeding, veterinary costs, logistics, and marketing costs, different reagents) depending on the herd size. The products are composed of milk, culled camels, and eventually by-products such as wool or manure. Thus, the expected financial incomes as the difference between products and charges can be evaluated for the coming years, and the technico-economic model can be used as dialog tool with breeders.

Keywords

Animal identification · Camel demography · Farm management · Zootechnical performances · Incomes

Breeding of large camelids for milk production is aimed at sufficient profitability. However, it is confronted with the biological limits of the animal, whose productive cycle is slow, the survival of the young sometimes problematic and productivity low, even if this is partly compensated by its longevity and the good economic valuation of its products, milk particularly. It is therefore necessary to verify the economic

B. Faye et al., *Large Camel Farming*, https://doi.org/10.1007/978-94-024-2237-5_7

potential of such a farm. In this section, we will look at a few aspects of economic management, starting with (i) identification of the animals, an essential criterion to ensure individualized monitoring; (ii) demographic modeling in relation to technical performance; and (iii) economic modeling.

7.1 Animal Identification

In most countries where large camelids are bred, there is no standardized system for identifying animals, which immediately raises the problem of their exhaustive census. At the very least, it is advisable to clearly identify all dromedaries and Bactrians on a farm, especially dairy ones. This provides a means of monitoring performance, setting up a selection system by recording parentage, and possibly feeding a database of herd performances.

7.1.1 Types of Identification

Three types of identification are proposed:

- Ear tagging, probably the most common method, although not the most appropriate. The ears of large camelids are small, and it is not easy to place the ear tags properly; ear tag loss and ear tearing are also common, and the risk of infection is not negligible (Fig. 7.1). On the other hand, reading can be laborious, as the figures can be difficult to read from a distance, and this often requires immobilization of the head, thus causing stress for the animal. Finally, the numbers often disappear over time and need to be rewritten regularly. It is therefore an inexpensive technique, but not very efficient.
- Subcutaneous implants (electronic chips), a common practice in the Middle East. The method consists of placing a subcutaneous electronic implant that can be read by a suitable reader (http://www.cliniqueveterinaire-adk.be/la-puceelectronique/). In general, the implant is placed at the base of the neck or in the shoulder and can be read from a distance of a few dozen centimeters, thus not requiring the animal to be restrained. Its disadvantage is that it can migrate under the skin and thus be more difficult to locate. It can cause local irritation. The equipment includes a syringe for injecting microchips and a reader.
- Electronic boluses, the most relevant method in terms of efficiency and durability. It requires implantation via an esophageal probe in order to place the bolus in the rumen (compartment C1) where it will remain for the life of the animal. The reading is done with a rod connected to a reader, which makes it possible to remain at a distance from the animal to be identified (Fig. 7.2). However, the device is quite expensive.

For more details, one may usefully refer to the article by Caja et al. (2016).

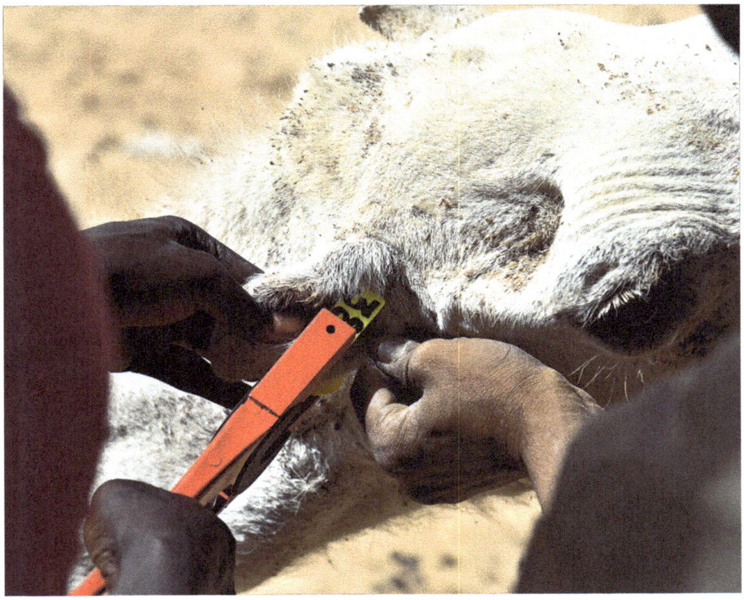

Fig. 7.1 Ear tagging on a dairy camel farm (Mauritania)

7.1.2 Implementation of a Farm Database

Animal identification is the key for implementing a dairy farm database. The purpose of a database is to follow up performances, production, and demographic changes. There is no typical database model. The construction of the database can be done under Access© and is based on a conceptual data model (CDM), which consists in establishing the list of parameters to be measured and kept as well as the links between all these parameters, as in the example in Fig. 7.3.

From the MCD, we can build the data sheets to be fed (in the form of linked Excel tables) with easy-to-use interfaces (Fig. 7.4).

The database monitors performance through various graphs that allow better management of the farm, for example, monitoring of daily milk production (Fig. 7.5).

7.2 Demographic Model

The demography of a herd on a farm is estimated based on entries (births and purchases) and exits (deaths, culling and sales), which in turn depend on a number of zootechnical parameters such as fertility rate for entries, mortality, abortion, offsprings, and culling rate for exits (Alzuraiq et al. 2015). Thus, the probability of moving from one age class to another within a herd (e.g., from 0–1-year-old female

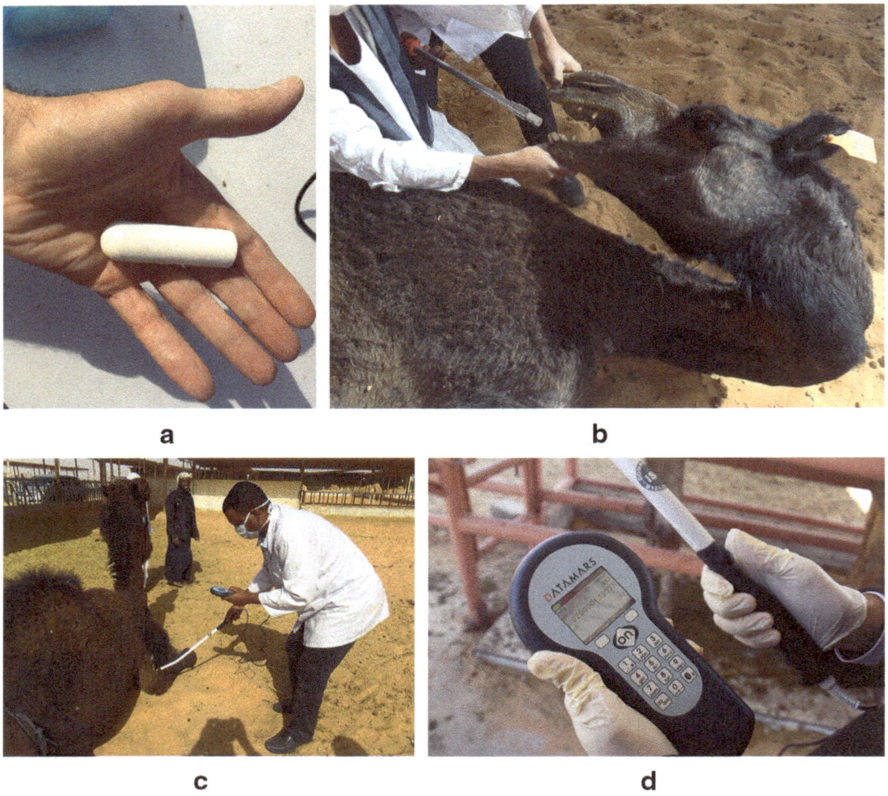

Fig. 7.2 Electronic bolus for identification of large camelids. (**a**) Datamars© device; (**b**) implantation of the bolus with an esophageal probe; (**c**) reading with a rod connected to the electronic reader; (**d**) reading device

camel to 1–2-year-old female camel) will depend on the mortality rate in the 0–1-year-old age class. With a rate of 10%, the probability of transition will be 0.90. The conceptual scheme of the demographic model is therefore summarized in Fig. 7.6.

The zootechnical parameters for estimating the population growth of a camel herd are described in Table 7.1:

By comparing the most unfavorable zootechnical performances (lower limits of the values listed in Table 7.1) with the most favorable performances (upper limits), one can test the hypotheses of demographic evolution (Fig. 7.7) or milk production (Fig. 7.8).

The demographic model is the basis for determining the economic profitability of dairy farming.

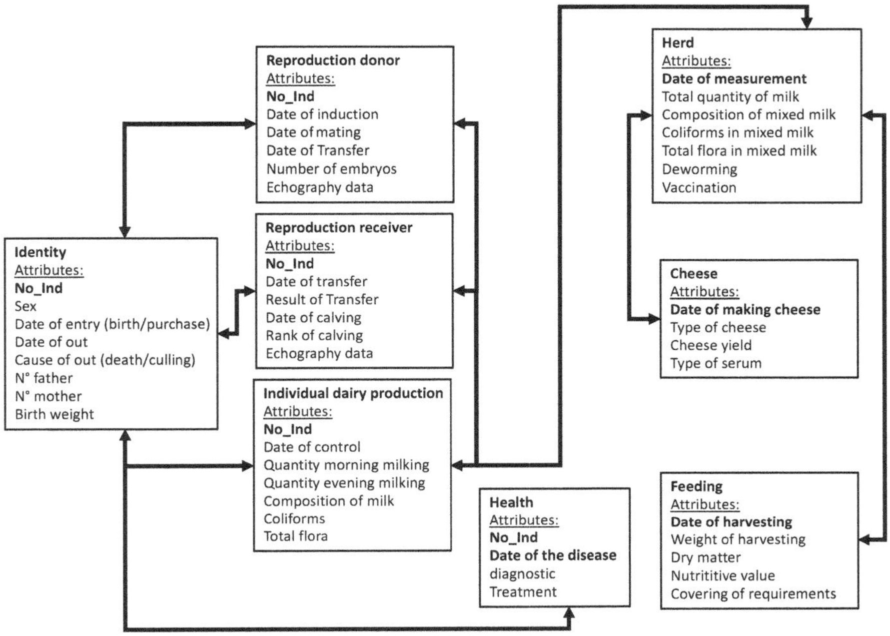

Fig. 7.3 Example of a CDM of the experimental dairy farm of the Mauritanian Centre for the Development of Camel Breeding (CMDEC) in Mauritania

7.3 Economic Profitability Assessment Model

In order to assess economic profitability, it is necessary to distinguish between expenditure items and products.

7.3.1 Expenditure Items

Regarding the fixed or variable expenses, the following parameters must be integrated:

- Cost of food
- Cost of veterinary care
- Manpower cost
- Marketing expenses
- Cost of breeding equipment
- Cost of fluids (water, electricity) and consumables (cleaning products)
- Cost of purchasing new animals
- Taxes and duties

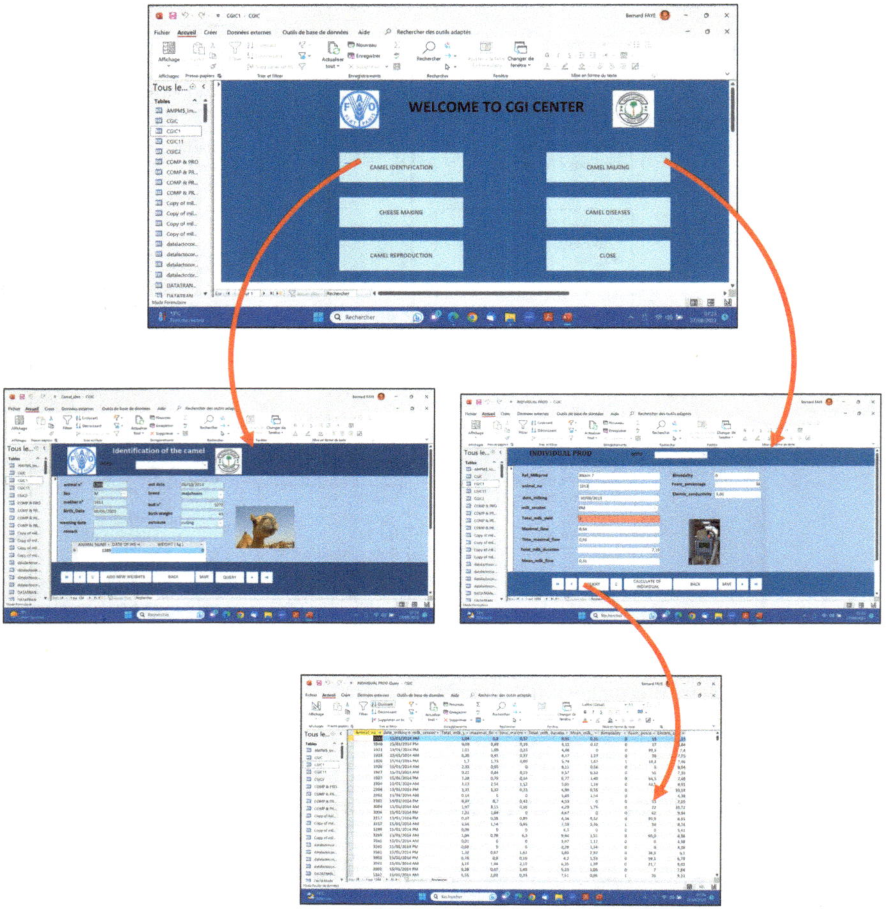

Fig. 7.4 A representation of some screens of the database of the Kharj experimental dairy camel farm (Saudi Arabia), showing (1) the home page, (2) the animal identification data sheet, (3) the milk performance data sheet, and (4) the final table of milk performance data

- Bank fees
- Depreciation costs for infrastructure and equipment

Some aspects and estimation methods are specified below.

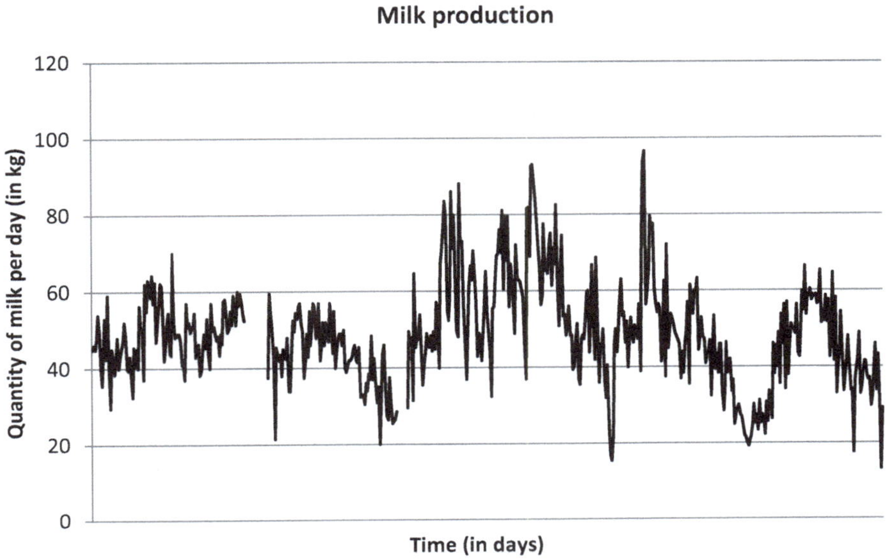

Fig. 7.5 Monitoring of daily milk production in Kharj dairy camel farm (Saudi Arabia) for 18 months from a database

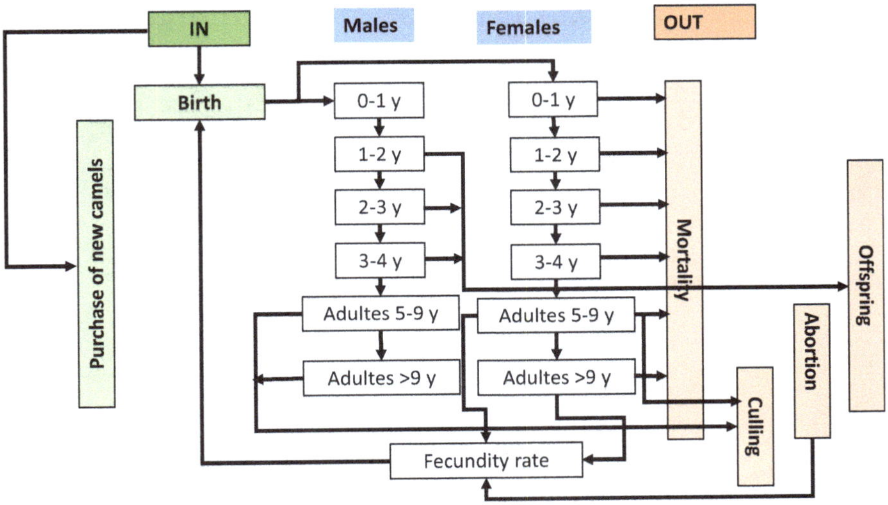

Fig. 7.6 Demographic model of a dairy herd

7.3.1.1 The Cost of Food

The cost of feeding depends on whether the animals have access to free-range or rented pasture. In the latter case, the cost of the land (land taxes, maintenance, rental, or amortization of the acquisition) must be integrated. In all cases, the amount of feed

Table 7.1 Input and output parameters of the demographic model (some cells are automatically calculated by their inverse: sterility rate for the fertility rate; female *sex ratio* for the male *sex ratio*)

Model input parameters	Values	Model output parameters
Annual infertility rate (%)	5–15	Total milk production (l)
Annual fertility rate (%)	85–95	Number of livestock (n)
Survival rate 0–1 year (%)	80–95	Rate of lactating females (%)
Survival rate 1–2 years (%)	90–95	Pregnant female rate (%)
Survival rate 2–3 years (%)	90–98	Rate of nonpregnant females (%)
Adult male survival rate (%)	95–100	Offspring rate (%)
Adult female survival rate (%)	95–100	Herd size and composition charts
Abortion rate (%)	5–15	
Farrowing rate (%)	85–95	
Male sex ratio	45–55	
Female sex ratio	45–55	
Adult male culling rate (%)	5–10	
Female cull rate 5–9 years (%)	5–10	
Culling rate for females >9 years (%)	20–30	
Culling rate for young males (%)	80–95	
Length of lactation (days)	300–330	
Progression of milk productivity (%)	0–5	
Milk production/day/animal (l)	5–10	

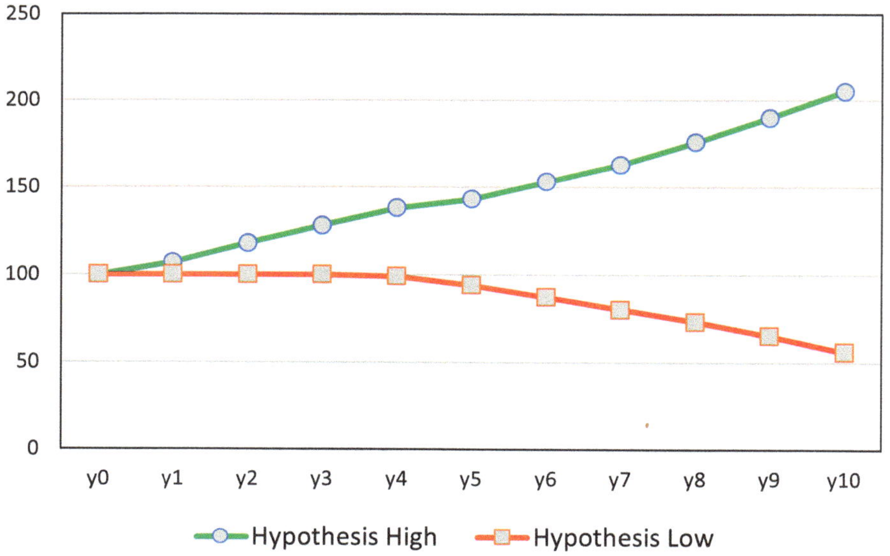

Fig. 7.7 Theoretical evolution of the camel population (based on 100 animals of which 36% are lactating females) over 10 years according to minimum (red curve) or maximum (green curve) zootechnical performances

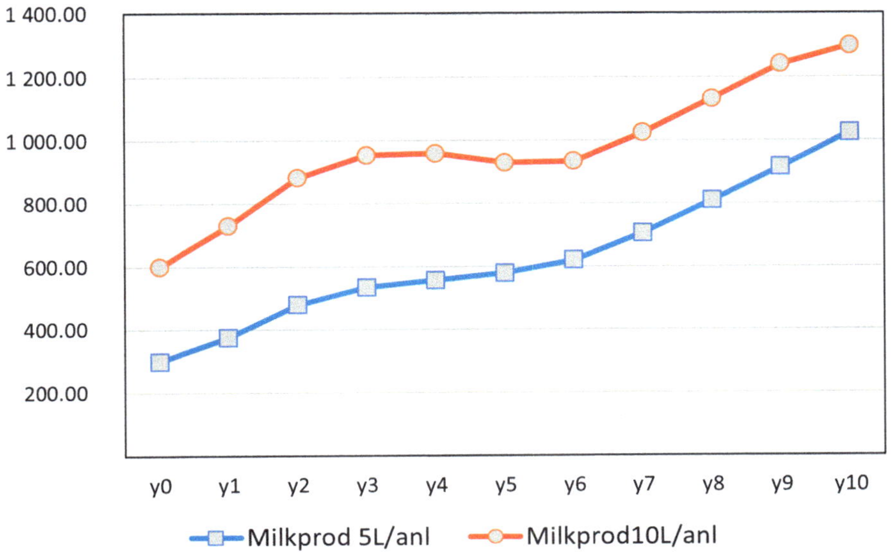

Fig. 7.8 Evolution of milk production over 10 years according to the average individual productivity of 5 l/d (blue curve) or 10 l/d (red curve) for optimal zootechnical performances

Table 7.2 Estimated quantity of fodder and concentrates by weight and status of camels for the calculation of feed costs

Category	Average weight (kg)	Forage quantity (kg DM)	Quantity of concentrates (kg)
Camel 0–1 year	80	1.6	1
Camel 1–2 years	160	3.2	2
Camel 2–3 years	250	5	2.5
Prepubescent	450	9	3
Adult male	700	14	1
Adult female	550	11	3

required must be estimated according to the composition of the herd (Table 7.2). The cost of feed will be estimated by multiplying the estimated quantities by the price of fodder (if purchased and not produced on the farm) and concentrates.

7.3.1.2 Veterinary Costs

Estimating health-related costs is difficult because there may be unforeseen expenses. There are two types of expenditure: (i) preventive expenditure (vaccination, deworming, veterinary pharmacy) and (ii) veterinary intervention expenditure. It is possible to estimate them using the concept of the veterinary intervention unit

(VIU), which has been developed by the veterinary groups under agreement. The principle consists of establishing number of VIUs per category of animal. By default, 1 VIU is allocated for camels under 1 year of age, as there are more interventions in this age group, compared with 0.2 VIU for other age categories up to puberty, 0.1 VIU for adult males, 0.3 VIU for pregnant camels, and 0.5 VIU for lactating camels. This allows the total number of VIUs to be calculated for the whole herd according to its composition. The cost of a VIU is then estimated, i.e., the average expenditure for camels under 1 year of age (e.g., 50 or 100 €/year).

7.3.1.3 Manpower Cost

The manpower (skill level and number) comprises a fixed number of people for administrative and management tasks and a variable number for technical tasks. For the latter, the number depends on the size of the herd. For machine milking, there is usually 1 milker for every 25–30 milking camels, 1 shepherd or animal handler for every 60–100 animals, and 1 animal technician for every 120–150 animals. For hand milking, one milker is needed for about 20 camels.

7.3.1.4 Depreciation Costs

In general, infrastructure is depreciated over 20 years, i.e., 5% per year of the price of the buildings and equipment (vehicles, milking equipment, possibly a control laboratory, milk tank, office equipment) over 5 or 10 years depending on the type, i.e., 10% or 20% per year of the investment cost.

7.3.1.5 Expenses Related to the Marketing of Milk

This expense item depends on whether it is a bulk sale to a dairy or a direct sale to individuals. Expenses include transport costs and packaging costs.

In total, and based on a farm of about 100 camels, the share of expenses related to feed represents between 30% and 40%, while those related to manpower vary between 40% and 50%. These two items of expenditure account for between 60% and 80% of expenses (Fig. 7.9).

7.3.2 The Products

They are of two kinds: (i) milk production, the quantity of which depends on the number of lactating camels and their productivity, and (ii) culled animals and sales of animals for breeding or slaughter. Concerning milk, an added value can be obtained in case of on-farm processing (cheese, fermented and pasteurized milk), but this implies additional investments in equipment. On a dairy farm, the milk product represents most of the income (about 90–95%) and varies depending on whether the male camels for sale are fattened on the farm or sold to external fatteners (Fig. 7.10).

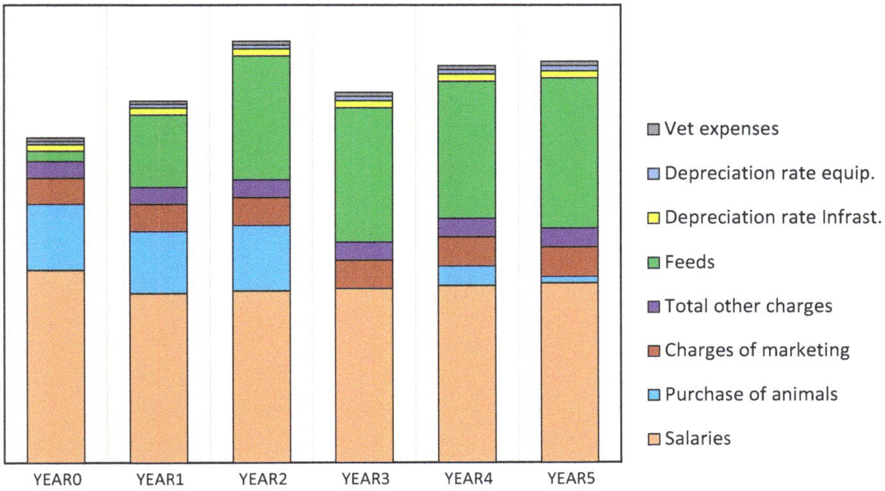

Fig. 7.9 Load structure on a dairy camel farm for 5 years

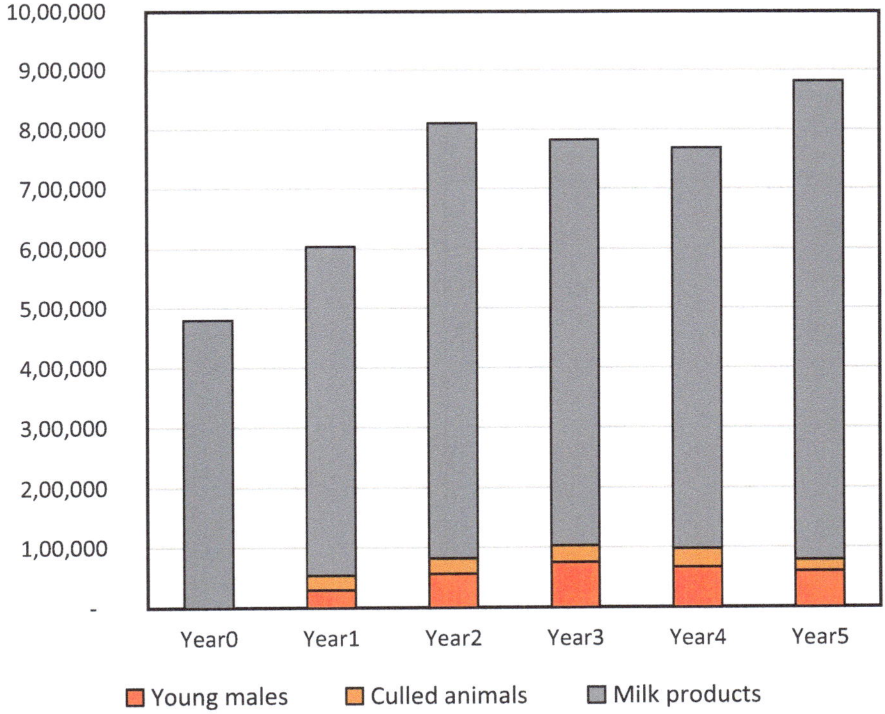

Fig. 7.10 Product structure of a dairy camel farm for 5 years

7.3.3 Final Profitability

The evaluation of the profitability of a dairy farm will therefore depend on many parameters, both zootechnical (productivity, performance) and human (management) as well as initial investment. The use of the demographic model combined with the evaluation of costs and products allows the expected financial results to be visualized (Fig. 7.11). The profitability can also be assessed by calculating the "internal rate of return" (IRR) according to the formula:

$$\frac{\sqrt[n]{CF_n}}{I} - 1$$

where CF is representing the cash flow occurring at time n, and I is reflecting the initial investment (animals, infrastructures, and equipment). The cash flow is the amount of cash coming and going out from the farm: milk and animal selling, other services. Another parameter to appreciate the profitability is the "payback period for capital," i.e., the ratio initial investment/annual cash flow allowing to assess the number of years necessary to reimburse the capital. This parameter is also strongly dependent on the total initial investment, volume and price of milk, number and price of culled animals, and percentage of expected margin.

The demographic model associated with the evaluation of expected expenses and incomes according to zootechnical performances constitutes a tool for dialogue and therefore for forward-looking management. It enables different scenarios to be tested

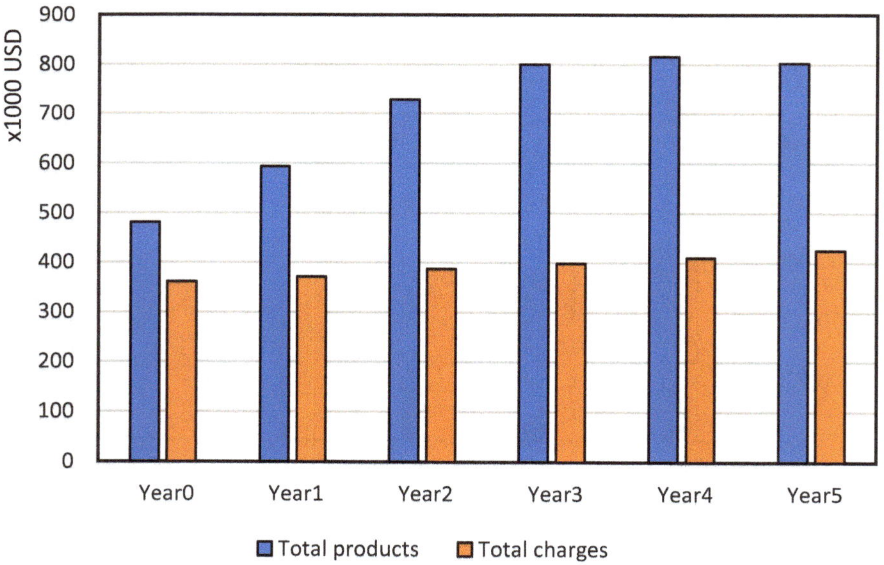

Fig. 7.11 Total income and expenses for 5 years in a dairy camel farm with an average productivity of 5 l/animal and average zootechnical performance

and, above all, priority actions to be determined by answering questions such as the following: Is it better to improve the fertility rate or reduce the mortality rate of young animals as a priority in order to improve profitability?

References

Alzuraiq F, Faye B, Lesnoff M (2015) Use of demographic model to assess the potential change in camel population and economy: the example of Saudi Arabia. In: Proc. of 4th conference of ISOCARD, "silk road camel: the camelids, main stakes for sustainable development", June 8–12, 2015 Almaty, Kazakhstan, Konuspayeva G. (Ed.), special issue of Scientific and Practical Journal Veterinariya #2 (42) 2015, oral communication, 312–314

Caja G, Díaz-Medina E, Salama AA, Salama OA, El-Shafie MH, El-Metwaly HA, Ayadi M, Aljumaah RS, Alshaikh MA, Yahyahoui MH, Seddik MM, Hammadi M, Khorchani T, Amann O, Cabrera S (2016) Comparison of visual and electronic devices for individual identification of dromedary camels under different farming conditions. J Anim Sci 94(8): 3561–3571. https://doi.org/10.2527/jas.2016-0472

Camel Products and Services: From Dairy, Meat, and Nonfood Products to Riding and Transport, Including Slaughter

8

Abstract

With the modernization of the camel sector, camel milk is processed to more diversified products than before. In addition to traditional fermented milk, pasteurized milk, powdered milk (for international market), cheese, butter, yogurt, ice cream, and various sweets appeared in the markets of most countries. The processing into cosmetics is also more and more popular. The camel meat sector is also seeing a certain modernization including better slaughtering conditions, though there are still those who, for religious purposes, continue to slaughter the traditional way with a knife. Besides traditional processing for allowing long-term storage (drying, smoking, brining), new processes for making sausage, ham, or terrines are now available. Nonfood products such as wool or skin are also experiencing greater development, thanks to different technical innovations in the wool and leather industry which is courting the international market. The use of camel manure remains marginal although urine has entered into traditional pharmacopeia. Camels are also essential for different services as riding (using different types of saddles) including racing and other sports, packing and carting for carrying a wide range of goods, and pulling equipment as a contributor of agriculture material, touristic events, or collective transport. Such high variety of services leads one to consider the camel as the most important multipurpose domestic animal.

Keywords

Milk processing · Meat processing · Multipurpose use · Camel sector modernization · Camel welfare

B. Faye et al., *Large Camel Farming*, https://doi.org/10.1007/978-94-024-2237-5_8

8.1 Camel Milk Processing

For a long time, camel milk consumption was limited to raw or fermented milk. With its entry into the national and even international markets, the product is now diversifying its range of products, making it possible to offer consumers a wider range of products, such as pasteurized or flavored milk, ice cream, cheese, and various sweets. Trials for the manufacture of butter and yogurt have also been tested. But between laboratory developments and small-scale or industrial production, what assessment can be made about the products currently available?

8.1.1 Fermented Milk: From Traditional Product to Dairy Industry

Fermentation used to be one of the only ways to prolong the consumption of fresh camel milk produced in desert regions where it was not possible to preserve it by other means, due to the difficulties of natural coagulation, notably to make cheese paste. There is a wide variety of traditional fermented products linked to the diversity of the lactic flora responsible for fermentation and to cultural processing habits: *shubat* in Kazakhstan, *gariss* in Sudan, *suusac* in Kenya, *laben* (or *lben*) in the Arab countries, ititu or *dhanaan* in Ethiopia, and *chal* in Iran and Turkmenistan (Konuspayeva et al. 2023). Other fermented drinks based on camel milk are offered in Mauritania and Morocco, such as *zrig*. More than fresh milk, fermented milk is sought after by informed consumers for its probiotic properties (thanks to the presence of lactic bacteria) and its effects on health in general.

Natural fermentation (generally activated by "old" milk already fermented at room temperature) in a rural environment is never standardized and results in a final product of highly variable organoleptic quality, which is not compatible with industrialization of the process. The dairy industry has therefore sought, beyond the identification of the lactic flora, to characterize the functional properties of this flora in order to control fermentation and standardize the final product. The aim was to "modernize" fermented milk, as in Kazakhstan or China, by finally obtaining a product with constant properties in terms of taste, quality, and production cost (Konuspayeva and Faye 2021). To achieve this, the steps consist of the following:

- Identify the lactic acid bacteria (LAB) of technological interest and the yeasts responsible for natural fermentation.
- Select the strains of bacteria and yeast according to the expected final product.
- Test the growth kinetics, acidification capabilities, organoleptic properties, and antagonistic activities of each selected strain.
- Test nutrient media for optimal growth in industrial bioreactors.
- Prepare a special package of freeze-dried LAB (Fig. 8.1) to be used as standard ferments for milk producers.

Some countries are marketing *kefir* or *laben* with a controlled fermentation process, as in Algeria (Fig. 8.2), according to a methodology developed in Saudi

Fig. 8.1 Freeze-dried ferments for the preparation of *shubat*, fermented camel milk, in Kazakhstan

Arabia (Algruin and Konuspayeva 2015). However, in this case, the lactic ferments are isolated from cow's milk.

8.1.2 Pasteurized Camel Milk

Many countries in Africa and Asia have introduced pasteurized camel milk, which remains the most common "modern" processing. However, the technology used is most often based on the knowledge gained from cow's milk. Yet, the technical standards used for camel milk are far from being mastered, and a great technological variability is observed. It is also notable that the indicators generally used to control pasteurization (alkaline phosphatases) are not applicable to camel milk, given their heat-resistant properties in camels. Some authors have therefore proposed that alkaline phosphatases be replaced by γ-glutamate transferase (GGT) or lactoperoxidase (LP) still present in pasteurized milk (Wernery et al. 2008). However, pasteurization, although it contributes greatly to the reduction of microbial loads, is not equivalent to sterilization. Pasteurization can divide by 100 the microbial load, but with a highly contaminated milk (e.g. one million coliforms/ml), the remaining microbial load remains high (10,000 coliforms/ml while the standard is less than 100/ml). It is therefore necessary to be cautious about the initial quality of the milk. It does not allow to abstain from the need to ensure the production of milk with good hygienic quality. Also, the question of sterilization arises regularly, given the frequent poor hygienic quality of camel milk collected in pastoral areas.

Fig. 8.2 Camel milk *laben* from the el-Oued dairy (Algeria)

Moreover, in terms of equipment, it is advisable to prefer continuous pasteurizers (in tube) rather than pasteurizers with plates. Indeed, the fat in camel milk, the composition of which is different from cow's milk, tends to clog the plates during heating.

8.1.3 Sterilized Milk

The solution could be UHT milk. Unfortunately, the behavior of camel milk proteins at high temperature is a strong constraint for sterilization processes. Their denaturation prevents until nowadays to offer UHT camel milk, as proteins become unstable at high temperature. Many studies have tried to find a solution, but so far without success. After UHT treatment, camel milk shows a two-phase separation due to protein precipitation, which makes it commercially unsuitable for consumption. In fact, the serum proteins are rapidly denatured at high temperatures.

However, camel milk can be sterilized after reconstitution from milk powder as practiced by the Camelicious dairy in Dubai. To sterilize liquid milk directly, two technological solutions have been proposed (see the review of Konuspayeva and Faye 2021):

- Sterilization by high hydrostatic pressure, a technique that does not require heating and therefore avoids thermal denaturation of milk proteins. This method has been tested on camel milk on an experimental basis. With pressures of about 200 MPa for 5 minutes at 40 °C, the bacterial contamination of milk is significantly reduced without changing its composition. However, with higher pressures (above 400 MPa), denaturation of proteins is observed again, especially of α-lactalbumin.
- Microfiltration, a technique that preserves all the properties of the milk while increasing its shelf life to 1 month. However, only experimental trials have been conducted with camel milk. In the trial mentioned, the microbial load became insignificant, decreasing by 99% in the permeate. This microfiltered milk was found to be bacteriologically stable for 60 days without any change in the milk composition.

Thus, sterilization of camel milk is not yet really on the agenda. The promising technique of microfiltration requires working with skimmed camel milk, as its fat tends to clog the filters quickly.

8.1.4 Camel Milk Yogurt

There is an abundance of literature mentioning the possibility of making yogurt with camel milk. Several strains of classical lactic acid bacteria have been tested such as *Lactobacillus bulgaricus* or *Streptococcus thermophilus*, but also *Lactobacillus acidophilus*, *Lactobacillus casei*, and *Bifidobacterium*.

However, the manufacture of camel milk yogurt poses a problem of texture, as the product appears sticky and ultimately unpleasant to the palate. Indeed, the viscosity of the product does not change during the gelation process compared to milk from other dairy species. This constraint is of course related to the protein composition of camel milk or even, according to some, to antibacterial factors naturally present in camel milk. To obtain a better texture, trials with the addition of gelatin, alginate, or calcium have been attempted or with ferments producing exopolysaccharides. The application of a high-pressure treatment could have a positive effect on texture, but no trials have been conducted to date with camel milk.

Other authors have tried to improve the manufacture of camel milk yogurt by mixing it with that of other species or by introducing 0.75% biosynthesized xanthan, but with mixed results in terms of organoleptic properties. However, in all cases, the final product is at best a "drinking yogurt" without the taste qualities, even with the addition of natural or synthetic flavors. These difficulties explain why there is no real industrial production of camel milk yogurt at present, and the numerous trials have hardly left the laboratory. Some researchers have proposed frozen yogurt to obtain a product that is somewhere between yogurt and ice cream. The optimal composition from the point of view of texture would be possible with several ingredients such as fat (5%), sugar (13%), gelatin (0.5%), and banana (14%), but here again, such a proposal has never left the laboratories (Konuspayeva and Faye 2021).

8.1.5 Camel Butter

The fat in camel milk contains less than 0.5% butyric acid, compared to nearly 5% in cow's milk (Konuspayeva et al. 2008). In addition, the fat globules are smaller than in cow's milk. Consequently, the butter yield is low and the organoleptic properties disappointing. To obtain fat globules at the time of manufacture of butter, it is necessary to proceed with vigorous shaking hot (22–23 °C), which allows to recover about 80% of the fat (Berhe et al. 2013).

Ghee (clarified butter), a popular product in India, has also been attempted from camel milk, but in addition to the very low yield compared to buffalo or cow milk, the final product was found to be more susceptible to rancidity. The transformation into butter does not seem fundamentally interesting in the context of an industrial valorization of camel milk. In fact, apart from the trials carried out in Ethiopia, where the consumption of butter, including rancid butter for certain recipes, is high, the manufacture of camel butter has little future.

8.1.6 Camel Cheese

The difficulty of obtaining a good quality curd has always been the main reason for the lack of cheese processing of camel milk. Its low concentration in κ-casein, responsible for coagulum quality, is the main reason. In addition, the slow acidification of camel milk due to its stronger natural buffering capacity must also be considered. It is only recently that these technological difficulties have been resolved. After initial attempts in the late 1980s to improve the coagulation process by adding calcium to bovine rennet, which led to low cheese yields, and after numerous attempts to improve the process by using vegetal coagulants with mild success, the coagulation problem was solved, thanks to the work of Kappeler et al. (2006) on the structure of camel chymosin. By genetic transfer of the camel chymosin synthesis gene to the mold *Aspergillus niger*, the recombinant enzyme could be industrially produced and then marketed under the name Chymax-M1000® by Ch. Hansen© (Denmark). However, except for Mauritania, where Caravan® cheese was marketed by the Tiviski dairy in Nouakchott, and some limited trials in Morocco (Fig. 8.3), India, Saudi Arabia, Kenya, and the Gulf States, camel cheese has rarely left the laboratory.

Different technologies have been tested since then to produce various types of cheese. For example, technologies have been used to make gruyere (Fig. 8.4a); mozzarella, feta, and halloumi (Fig. 8.4b); and processed cheese or cottage cheese (Konuspayeva et al. 2017; Konuspayeva 2020).

However, the manufacture of camel cheese faces three pitfalls:

- Final result does not correspond to the expected appellation (the taste of camel "gruyere" has nothing to do with that of real gruyere).
- Acceptability to local consumers is highly variable, as the final product does not necessarily match their taste experience and consumption habits.

Fig. 8.3 Ajban Dakhla cheese made from camel milk occasionally produced in Morocco

a b

Fig. 8.4 (**a**) Camel cheese made using Gruyere technology (Saudi Arabia). (**b**) Feta cheese preserved in olive oil (Saudi Arabia)

- Final cost of the product is high, even if the yield is comparable to that of cow's milk, due to the high price of the raw material.

To overcome these difficulties, several adjustments have been tried, such as the production of mixed cheeses, mixing camel milk with that of other species, the marketing of cheese that best corresponds to the tastes of local consumers, or the use of whey in the form of soft drinks.

One of the obstacles to the development of cheese production is the final price of the product. Indeed, as the raw material is more expensive than cow's milk (between two and ten times depending on the country), the price of camel cheese is becoming unaffordable for households in the South, especially since, until recently, we did not know what to do with the whey. It can be used to feed monogastric animals (especially pigs, which is unimaginable in most camel countries, which are Muslim) or in preparations such as ricotta. Several trials have been carried out with camel milk whey to help reduce the price of cheese (whey represents nearly 90% of the initial volume of milk). This is even more interesting because camel whey is richer than cow whey. It contains more fat (9 g/l vs. 7.7 g/l for cows) and more protein (9.21 g/l vs. 7.3 g/l) for the same dry matter content (Konuspayeva et al. 2008). Recently, in Mauritania, it has been possible to offer fresh or fermented drinks based on camel whey with excellent consumer acceptance.

8.1.7 Camel Milk Powder

The development of an international market for camel milk was only possible with the appearance of milk powder. This is an important issue because of the distance between the production and consumption areas. Once transformed into powder, transport only concerns the dry part of the product, i.e., between 11.5% and 12.5% of the weight. One ton of fresh milk corresponds to 120 kg of powder on average (Fig. 8.5).

To manufacture this powder, several technologies have been applied to camel milk: hot *spray-drying* or freeze-drying. The first method seems preferable for a better reconstitution of the liquid milk, but it implies a more important investment for the acquisition of a drying tower and a sprayer. On the other hand, the powder produced by freeze-drying could theoretically be easily used in the food industry (pastry, chocolate).

8.1.8 Other Camel Milk Products

The production of ice cream with different flavors is easy. There are several examples of commercialization, for example, in the United Arab Emirates, Morocco, and Kazakhstan (Fig. 8.6). Manufactured using the same technology as for other milks, ice cream is popular with consumers and, above all, with less consumer reluctance than with other camel milk products. However, very few studies on texture and sensory properties have been conducted.

There are no references on the processing of camel milk into sweets. However, there are some traditional products. For example, in Kazakhstan, a caramel called *Balkailmak* is obtained after a long heat treatment of about 10 hours at boiling

Fig. 8.5 Camel milk powder (bulk) sold on the Alibaba. (© platform in China)

Fig. 8.6 Camel milk ice cream marketed in Morocco

temperature. Another example is the introduction of milk powder in chocolate, as it is proposed in the Emirates.

8.1.9 Nonfood Processing of Camel Milk

The manufacture of soaps and other cosmetic creams with camel milk is now a common practice in many countries, either on a semi-industrial or artisanal scale. China markets kits containing a variety of cosmetic products as lipsticks, moisturizers, shampoos, and various lotions (Fig. 8.7). For the cosmetic industry, camel milk benefits from the hypoallergenic properties of its proteins.

The implications of these data for camel milk can be summarized in Table 8.1, which shows the opportunities, interests, and constraints for each potential product.

8.2 Camel Meat Products

The breeding of large camelids is also seen regarding the production of meat and meat products throughout the world. There are also reference books on this topic covering everything from slaughtering techniques to the nutritional qualities of large camelid meat (Kadim et al. 2013) that can be referred to. We will limit ourselves to a few salient points in the context of this book, especially since the slaughter of large camelids is not yet regulated in countries of new settlements such as in Europe.

Fig. 8.7 Box of camel milk cosmetics made in China

Table 8.1 Interests and constraints of the processing of various camel products

Product	Interest	Constraints		
		Technique	Economic	Social
Fermented milk	Easy to produce, requires little investment, and is well accepted socially (tradition of *leben* and *zrig*) Can be stored for several days or even weeks Proven probiotic qualities	Requires a good standardization of the organoleptic qualities of the product Transport with respect to the cold chain	Cost of transporting a liquid product	Must be consumed as is, but can be used in traditional culinary preparations (couscous)
Pasteurized milk	Allows for medium-term storage (1 week) Has an image of quality milk because it is heat treated	Requires appropriate equipment and good quality raw material Transport with respect to the cold chain Relatively limited shelf life	Cost of transport Investment costs (pasteurizer) and importance of packaging	Urban product by excellence
Milk sterilized by microfiltration	Allows for long-term storage (30 days)	Requires appropriate equipment (microfilter)	Capital cost, as a pasteurizer is required Cost of transport	Urban product
UHT milk	Allows for a very long shelf life (2–3 months)	Impossible with camel milk	Cost of transport	Urban product
Yogurt	Diversification of supply Many possible products (plain, sweet, flavored) Low investments	Low texture Requires raw material of good hygienic quality	Product to be well adapted to local consumers	Poor acceptance due to texture
Butter	Low interest	Low yield and poor organoleptic qualities	Requires a cold chain Low investment (churn)	Poor consumer acceptance
Cheese	Good diversification of the offer Low investment Low cost of transport	To make a product adapted to local consumers Requires extensive testing	High product price Narrow market	No strong cultural consumption habits Urban product

(continued)

Table 8.1 (continued)

Product	Interest	Constraints		
		Technique	Economic	Social
		Need to valorize the by-product (whey)		
Powder	Very long shelf life Possibility to export Agribusiness opportunities Low cost of transport	Choice of equipment for good quality powder Requires good technical skills (labor)	Requires significant investment (drying tower) Requires a large quantity of raw material to make the equipment profitable	Relatively few outlets for reconstituted milk at the national level Difficulty of access for the Western market (health barrier)
Ice cream	Very easy to make Little investment	No particular difficulty. Requires several flavors	Sensitive transport Local consumption	Choice of flavors
Sweets	Good product diversification	Requires fine-tuning	Collaboration with agri-food industries Requires a good knowledge of the market	Search for a good quality/price ratio
Cosmetics	Good product diversification Ease of export and transport	Choice of equipment according to the products	Many competitors (local crafts)	Request to aim for quality to meet the market

Unlike camel milk, which has only recently become the subject of a global trade, the trade in live camels for meat production has long been part of a flourishing regional trade with significant flows from Sahelian Africa to the Maghreb and especially from the Horn of Africa to the Arabian Peninsula.

8.2.1 Slaughter

In pastoral milieux, the slaughter of a camel for its meat remains rare, given the size of the animal which requires large number of guests. It is therefore limited to important festive events, such as the *gdaasiga* among the Afar pastoralists of Ethiopia and Djibouti.

The animals slaughtered are of all ages, but there are significant differences between the producing regions. In the Gulf countries, preference is given to male camels under 2 years of age, often imported from the Horn of Africa (Somalia,

Sudan, Djibouti, Ethiopia, Kenya), sometimes fattened specifically for meat production. In the Sahelian countries, slaughter mainly concerns adult animals (5–7 years). In some countries, it is forbidden to slaughter animals under a certain weight (e.g., 250 kg in Tunisia) or adult females (Saudi Arabia), except for old, culled females. These differences obviously have an impact on the quality of the meat.

8.2.1.1 Type of Slaughter

Slaughter conditions are often disastrous in most camel countries. This concerns not only the rules of hygiene (equipped slaughter area or slaughterhouse, specific clothing for slaughterers, suspension of carcasses, cold room, etc.), but also the notions of well-being (transport, conditions of entry into the slaughter chain, preslaughter stunning) contributing to a killing in the best possible conditions. The traditional and modern slaughter methods can be described in their basic forms as follows:

The traditional method is to place the animal in a sternal decubitus position (sitting position), bend the neck over the chest, and slice the neck at its base near the chest with a sharp blow, cutting the carotid arteries cleanly to allow a sudden loss of blood volume resulting in rapid loss of consciousness. No prior stunning is performed. The animal is then butchered starting with the back (unlike most other species), due to the presence of the hump. The hump is then removed, allowing the carcass to be turned over onto the skin, which serves as a protection against the ground. It is only at this stage that the animal is eviscerated.

The modern method consists of stunning the animal beforehand (electrical stunning) and then slaughtering it, either in the position explained above or with the animal suspended (but unconscious) by a hind leg. It should be noted that electrical stunning is permitted for ritual slaughter. However, this practice is not universally applied, including in modern slaughterhouses, where the animal can be suspended while conscious, in total contradiction with the basic rules of animal welfare. In the case of suspension, evisceration can be carried out primarily through a ventral incision as for other species. The advantage of this is that evisceration is quicker after killing and the risk of contamination is reduced.

8.2.1.2 Carcass Cutting

There is no standard carcass cut for camelids, but various cuts have been proposed. The most common is to separate the forequarter from the hindquarter by cutting the rib cage between the 11^{e} and 12^{e} ribs (Fig. 8.8). The forequarter itself is divided into five pieces (neck, shoulder, breast, ribs, rib eye), while the hindquarter consists of three pieces (loin, flank, and thigh). Other cuts have been proposed (Hermann and Fisher 2004).

Carcass yield varies according to the type and age of the animals slaughtered, ranging from 45% to 55% with an average of 50%. It is higher in the Bactrian camel. The weight of the hump can vary considerably (from 3 to 100 kg) and can represent up to 25% of the carcass weight. For 100 kg of carcass, the average muscle weight is 56–57 kg, bone 20–25 kg, and fat 14–18 kg. The low proportion of fat is remarkable and is an essential characteristic of camel meat. Finally, unlike cattle, the forequarter

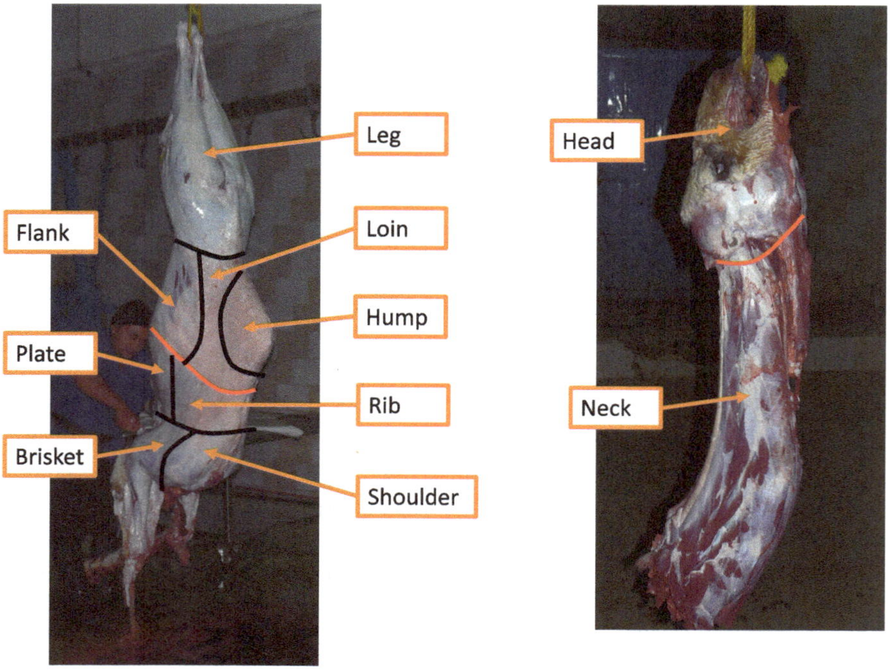

Fig. 8.8 Typical dromedary carcass cut. The red line on (**a**) marks the separation between forequarter and hindquarter, while on (**b**) it indicates the distinction between the neck and head

is predominant compared to the hindquarter, partly because of the presence of the hump and partly because of the greater development of the shoulders compared to the muscle mass of the pelvis. Even without including the hump, it is estimated that the forequarter accounts for approximately 24% of the carcass compared to just 21% for the hindquarter.

8.2.2 The Nutritional Quality of Camel Meat

Relatively well studied, the nutritional quality of camel meat is, like other species, dependent on the type of animal, its age, its sex, and rearing conditions (Faye et al. 2013). The average composition can be estimated as follows (according to various authors):

- Water: 70–77%
- Protein: 17–23% (dromedary) or 17–18% (Bactrian)
- Fat: from 1.4% (lean animals) to 10.5% (older fatty animals)
- Minerals: 1.1–1.5%, including 37–80% potassium depending on the muscle

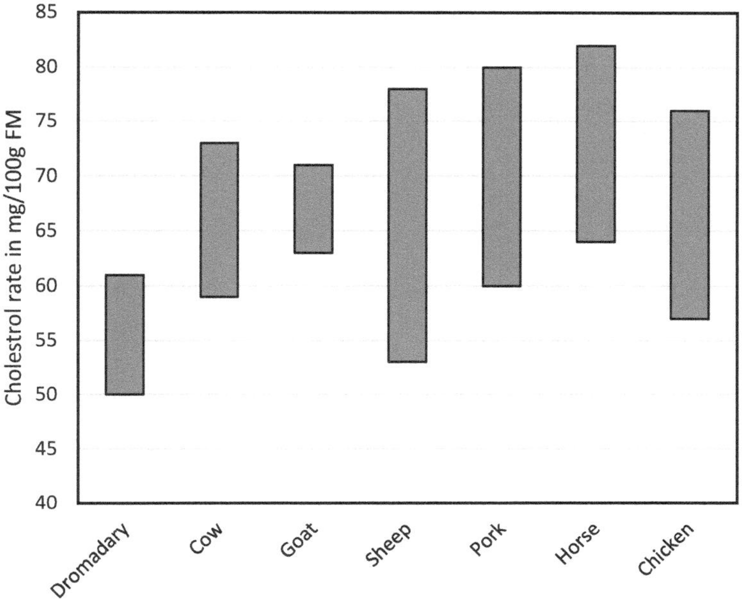

Fig. 8.9 Cholesterol levels in meat from several domestic species (mg/100 g fat [MG])

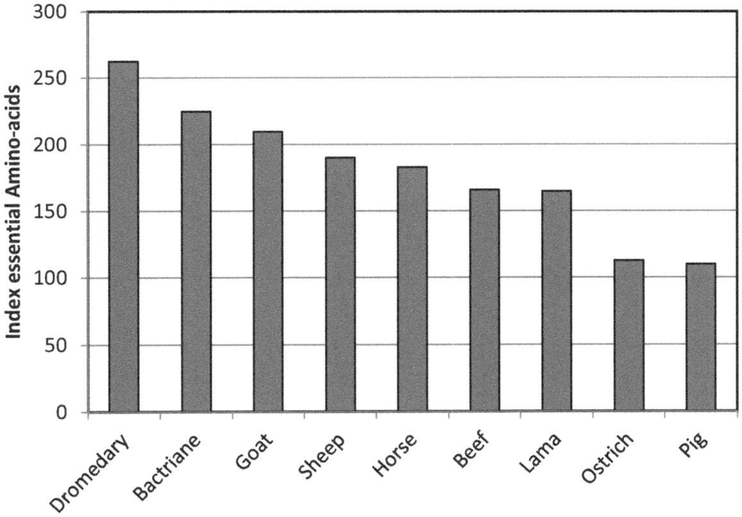

Fig. 8.10 Index of essential amino acids in the meats of different species most consumed by humans. (Source: after Raiymbek et al. 2015)

Overall, camel meat is known for its low-fat content (especially when the hump is removed from the carcass) and very low cholesterol, which makes it a potential commercial competitor to beef (Fig. 8.9).

The nutritional value of camel meat is also linked to the quality of its proteins. Indeed, one of the indicators used in human nutrition is the essential amino acid index (EAAI), which reflects the coverage rate of human needs. An index of 100 means that the product covers 100% of human needs. This index exceeds 250 in camel meat and reaches 225 in Bactrian camel meat. This is the highest index of all meats normally consumed by humans (Fig. 8.10).

In theory, a dromedary carcass would provide the energy needs of an adult man for 5 days and his protein needs for 35 days.

8.2.3 Processing of Meat Products

The processing of camel meat has recently undergone considerable development.

In traditional societies, the main method of preservation was to produce dried meat by solar drying. Indeed, when pastoralists sacrifice an animal, for self-consumption in the context of religious festivals, for example, and cannot use the entire carcass in a sufficiently short time, the only technical possibility is to obtain smoked meat, with added spices, which can be preserved for several months and which caravaneers appreciate during their travels. For example, among the pastoralists of the Horn of Africa, smoked meat goes by different names (*qwanta* in Ethiopia, *odka* in Somalia, *sharmoot* in Sudan and Chad), but the technology is

Fig. 8.11 *Khliâ* with olive oil in a shop in Rabat

Fig. 8.12 *Deve Sucuk* (camel sausage) at the Camel Wrestling Festival in Selçuk, Turkey

broadly the same: thin slices of meat are cut up and placed in a solar dryer or simply on wires in the sun. The dried product can be coated with spices or butter, recut, or ground into powder. In the Sahara (southern Morocco, Mauritania), several preparations based on traditional recipes are popular, such as *tichkar* (dried meat), *khliâ* (dried meat preserved in olive oil), or *loudek* (humpback fat). "Modernized" versions have been offered to consumers in large cities (Fig. 8.11). Thus, dried meat in various forms can be found in supermarkets, notably in the Gulf countries (*kadid*) or in the Sahelian countries (*kilishi*) or in East Africa (*biltong*).

Other practices such as smoking or brining are traditionally practiced.

In Turkey, where the camel herd is mainly used in "camel wrestling" festivals, the carcasses of culled animals, which are often relatively old, are processed into sausages (*deve sucuk*), which are presented in strings at festivals (Fig. 8.12) or in butcher shops. This product has recently been registered as a PDO (protected designation of origin) product.

In Middle Eastern fast-food restaurants, the *camelburger* has become a menu item. In China, vacuum-packed steamed camel legs (Fig. 8.13a), pâté, and terrines are available. In Kazakhstan, camel *kaze*, a sausage usually made from horse meat, is now offered to urban consumers (Fig. 8.13b). In Kazakhstan, sterilized, cooked preparations are also available (Fig. 8.13c), as well as *corned-camel* for the military.

In Morocco, a large processing company also offers a whole range of processed products based on camel meat: ham, sausage, mortadella, and luncheon meat (snack). Overall, camel meat lends itself very well to all the processing methods

a b c

Fig. 8.13 (**a**) Camel foot steamed (China). (**b**) Camel *kaze* (sausage). (**c**) Rice and camel meat dish in Kazakhstan

applied to other meats while relying on the local know-how and food habits of the countries concerned.

Several assessment tests by panels of tasters have been carried out to evaluate the physical and sensory qualities of camel meat, to assess the degree of acceptability of smoking or curing, or simply to evaluate the acceptability of the product compared to the same product prepared with beef. In all cases, the assessments appear favorable.

8.3 Other Camel Products

The large camelids are the very type of animal with multiple uses. In addition to the service functions of saddled riding or transporting goods or even, in the southern countries, agricultural work, we must add the production of milk and meat, as mentioned above, as well as wool, leather, and manure, all of which are described briefly in this chapter as a complement to the service functions.

8.3.1 Wool

While all camelids can produce a woolly fleece, it is much more abundant and of better quality in the Bactrian camel. The fleece is most abundant in winter, and a natural shedding of the woolly fleece takes place in late spring as the outside temperature increases. The fleece includes hair, wool in the strict sense, and down. Shearing, which is practiced in Turkmenistan and India, for example, allows the recovery of a fleece that is a mixture of hair and wool, which diminishes its quality, especially for making clothes. Also, the recovery of the wool by combing will be largely preferred as it is practiced in Mongolia or in China at the time of the spring molt. Indeed, this technique allows to recover only the noble part of the fleece.

a b

Fig. 8.14 Mechanical shearing of Arvana camels in Turkmenistan

Bactrian wool fiber is renowned for its insulating power (it was used to make underwear for Soviet cosmonauts!), its resistance, and its fineness (Hasi et al. 2020). A male Bactrian provides between 3 and 5 kg of fleece, but record figures of 18 kg are reported in the literature. In the dromedary, wool production is lower (about 1 kg) and shearing is not widely practiced, except in Central Asia (Fig. 8.14). Camel wool is a relatively fine fiber (14–30 μm in diameter), which can reach 40–50 cm in length for the longest fibers (especially on the shoulders and chest). In Europe, the use of camel wool remains very anecdotal.

While camel wool, which is of lesser quality, is traditionally used for the manufacture of carpets and tent velum, or even coats (*gandoura*), Bactrian wool is nowadays used for high quality blankets and clothing. In fact, if camel fiber is integrated into national textile industries, as in Tunisia and India, Bactrian wool, which has a cashmere quality, is highly prized on the international luxury market, a card that Mongolia and China have not failed to play.

8.3.2 Camel Leather

Dromedary leather has long been of poor quality despite its weight (between 22 and 47 kg) and size. It is indeed of low commercial value. The presence of the hump, the frequency of skin diseases such as mange or ringworm, and the crude butchering techniques (by the back as mentioned above in the section "Slaughter") add to the depreciation of the product. Traditionally, camel leather was used for saddlery and thongs because of its strength.

However, recent technological innovations (better fleshing, storage of skins in freezers, thinning of the dermis, better tanning techniques, treatment of skin diseases) have made it possible to obtain a better raw material that is now used in

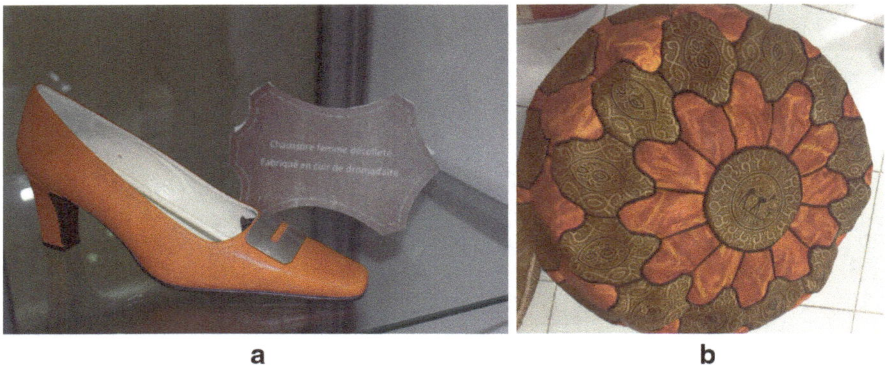

Fig. 8.15 (**a**) Camel leather shoe made in Tunisia. (**b**) Dromedary leather pouf made in Morocco

Fig. 8.16 Collection of camel dung at the Pushkar fair (India)

the composition of handicrafts (poufs, bags) and even luxury products (shoes, coats) distributed in the tourist industry, particularly in North Africa (Fig. 8.15).

For completeness, pastoralists also use camel sinew to make strong straps for tents or packs, and occasionally long bones can make useful tent poles.

Fig. 8.17 Paper made from camel dung in India. (Photo https://camelcharisma.files.wordpress.com/)

8.3.3 Manure

The ability of large camelids to recycle nitrogen from protein metabolism results in the excretion of urine low in ammonia and urea and feces low in nitrogen, therefore providing manure of relatively low fertilizer value. Furthermore, with a lifestyle largely based on herd mobility, the collection of excreta is a difficult operation, only permitted when there is a high concentration of animals at peculiar events such as the Pushkar fair in India (Fig. 8.16).

On the other hand, in intensive systems with sedentary animals, and especially in straw-bedded pens, manure harvesting is quite possible and can be used for market gardening. However, in most cases, camel dung is not very hydrated, so it dries easily and is used as fuel. Incidentally, in India it is used in the manufacture of paper (Fig. 8.17).

Incidentally, the dromedary droppings, well-dried, make excellent pawns for games in the sand, such as *Awale*, appreciated by Tuareg children. Finally, it should be noted that urine is sometimes collected for traditional pharmacopoeia. There are many beliefs about the effectiveness of urinotherapy in treating many ailments. Recent studies seem to indicate beneficial effects on cancerous cell lines, skin diseases (for external use), and certain parasites (for internal use). Regular consumption of camel urine would have hepato-protective effects. However, caution is needed to clearly distinguish (as for camel milk) what are empirical observations, attested therapeutic effects, placebo effects, or legends. It is likely that certain molecules from desert plants, eliminated in the urine, are responsible for the "therapeutic" effects attributed to camel urine, which is also relatively low in urea, the main toxic element.

8.4 The Service Functions

The use of large camelids goes beyond the sole function of production. We will mention here some of the uses of the animal throughout the world in other functions that are more related to services, whether through sport, transport, or agricultural work.

8.4.1 Saddled Mounts

The use of large camelids for saddled riding is probably as old as domestication, as evidenced by petroglyphs in Central Asia or rock engravings in Saudi Arabia (Fig. 8.18a and b), either for the movement of individuals for personal or commercial reasons in desert areas or for military purposes.

Sumerian frescoes dating from 2000 BC mention raids from what is now northern Saudi Arabia toward the valleys of the Tigris and Euphrates rivers by Arab tribes (under the name of *Alhamu*) riding camels (Fig. 8.19). This military vocation is perpetuated through the mehari companies and other *camel corps*, maintained in desert countries, even if the advent of 4 × 4 vehicles tends to limit the interest of such units in modern conflicts.

Saddled riding of large camelids is universal but remains more or less important depending on the pastoral societies. It is very common among the Tuareg and the Moors, but less common among the Afar, for example. In addition to these riding traditions, which vary from one country to another, there is a variety of saddle types depending on their position in relation to the hump. Overall, in the dromedary, four types of saddled riding can be distinguished (notwithstanding the variability in the shape of the saddles):

(i) Riding in front of the hump, with the camel driver resting his feet on the animal's neck (this is the case in particular of riding among the Tuaregs with the pommel of their saddle in the shape of a cross, which is very characteristic)
(ii) Riding on the hump with a saddle in the shape of a bowl, with the camel driver's legs spread out on either side of the chest (case of the Moorish saddle or the Sudanese saddle)
(iii) Riding at the back of the hump, with a saddle that is often rudimentary, but may have stirrups, used in particular by jockeys in races (the case of the Gulf Arabian saddle)
(iv) and of course in the Bactrian, between the two humps (Fig. 8.20)

Fig. 8.18 Saddled Mounts in camel: (**a**) Petroglyphs showing antelope hunting on Bactrian camels (Alashan Desert, Inner Mongolia, China), (**b**) Rock graving showing warrior mounted on dromedary camel (Hegra, Saudi Arabia)

Fig. 8.19 Sumerian fresco showing an Arab camel raid in the second millennium BC (Dumat Al-Jandal Museum, Saudi Arabia)

Fig. 8.20 Types of saddled riding with saddle placement in dromedary. (**a**) Before the hump (Tuareg saddle). (**b**) Over the hump (Sudanese saddle). (**c**) Behind the hump (Arabian saddle). (**d**) Between the two humps in the Bactrian camel

Fig. 8.21 Dromedary-mounted palanquin for the bride in Algeria (**a**) and Sudan (**b**); installation for Tuareg women during festivities in Niger **c**C)

While saddled riding is traditionally reserved for men, special facilities are provided for young wives or for women in general during festivities or even during transhumance (Fig. 8.21). However, female camel drivers of the Royal Camel Corps in Oman ride camels in the same way as men (Fig. 8.22).

Fig. 8.22 Royal Camel Corps women in the Sultanate of Oman

From the point of view of its performance, the dromedary can cover 50–100 km/day on sandy ground at an average speed of 10–12 km/h. During historic war raids, highly trained animals were able to cover 150–200 km in a single day. The Camel Marathon of Douz, Tunisia, saw the swiftest camels finish after about 2 hours of running at a speed of 20 km/h, like Olympic champions. In the Gulf States, camel racing is very popular and has a thriving economy. The animals run races of about 10–15 km at speeds of 30–40 km/h.

For touring, several types of saddles are available that can carry several people. These range from a saddle that wraps around the hump to the front and back, giving a seat long enough to carry two or three people placed one behind the other, to a saddle with two side seats, as practiced in the Canary Islands (Fig. 8.23). A new type of saddle with greater consideration of welfare is proposed in France for carrying two people on the back of camel (Fig. 8.24).

While camel racing is universal, it is in the Gulf States that it is most formalized. It has spawned a thriving sports industry, including professional organizations for race administration and doping control. This industry also relies on applied research institutes, mobilizing the whole range of modern technologies (genetic selection of champions, biotechnology of reproduction, physiology of effort, specific nutrition, training methods, etc.). The best runners are capable of galloping over 10 km at an average speed of 34 km/h with peaks of 40 km/h. Races over longer distances (20 km) are sometimes organized in Sudan or Saudi Arabia. Since the use of child jockeys was discontinued, racing camels are ridden by remote-controlled robots (Fig. 8.25). The training of youngsters starts at the age of 3 and preferably involves lighter, more manageable female camels. The price of champions can reach

a b

Fig. 8.23 Devices for tourism. (**a**) With side seats (Canary Islands). (**b**) Longitudinal saddle for several people (India)

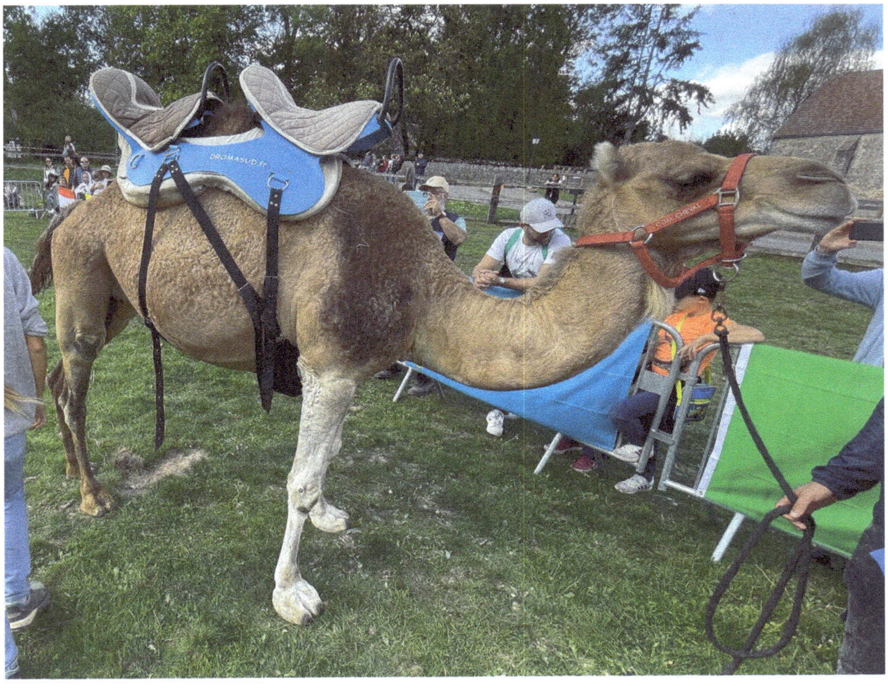

Fig. 8.24 Modern double saddle for tourism activity made in France. (©DromaSud)

considerable sums of money, millions of euros. An international federation of racing camels has recently been set up.

However, camel sport is not limited to racing. In Mongolia, the Bactrian camel is ridden in polo matches, for example.

Fig. 8.25 Racing camel in Saudi Arabia with its robot jockey equipped with an articulated arm carrying the whip

8.4.2 Packed or Hitched Transport

Before the industrial revolution and the motorization of travel, large camelids were the indispensable tool for transporting goods in desert countries. The caravans of the Silk Road or the trans-Saharan routes have shown throughout history the inescapable usefulness of these species for regional and even international trade, well before the globalization that is promoted today.

Large camelids are used in this context both as pack animals and as draught animals. Performance as a pack animal has been extensively studied. Thus, the dromedary can travel 40 or even 50 km/d at a speed of 4–5 km/h with loads varying from 100 to 300 kg depending on the morphology and weight of the animal, although records of more than 600 kg over very short distances have been reported, particularly with Asian breeds, which are heavier than African breeds. In Pakistan, competitions are held to see how far very heavy loads can be carried. Welfare considerations should, however, limit potential overloading. In colonial times, the regulations applied by the army's meharists companies limited loads to around 200 kg or less. The risks for the animal can be linked not only to the excess weight but also to the quality of the pack, as the wooden frames carrying the goods can injure the animal. Harnessing wounds are among the most common injuries suffered by pack camels.

The goods transported are highly variable, from specialized agricultural products (salt, cereals, dates) in the Sahara to manufactured goods (see the smuggling

Fig. 8.26 Moving the Somali pastoralists' dismantled hut in southern Ethiopia by camel

caravans carrying television sets in the Horn of Africa!), including books (*bibliocamel* for nomadic schools in Kenya and Pakistan), household effects during transhumance (Fig. 8.26), or rubbish collected on the beaches of the Atlantic in France.[1]

As with saddled riding, training for packing or pulling requires training from the age of 3–4 years, but the camel is not fully loaded until it is 6–8 years old, i.e., when it reaches its adult size. The career of a "carrier" animal can last 12 years.

Camel traction is widely used in South Asia (India, Pakistan), but also in several regions of Africa. In India camel carts, carrying everything that local trade needs to exchange (fodder, water, building materials, agricultural products) provide "eco-transport" (nonpolluting, economical, and self-sufficient energy) contributing to a truly sustainable development of the economy, both urban and rural (Fig. 8.27).

Numerous studies in India show remarkable performance, with a camel being able to pull 1.5–2 t on a two- or four-wheel cart (usually recycled aircraft wheels!) for 4 hours at 8–10 km/h. Overall, the pulling power is estimated at 2.8 kg/kg of live weight.

8.4.3 The Auxiliary of Agriculture

The service functions do not stop at the transport of people and goods but also include the animal as an agricultural auxiliary, contributing to systems based on the

[1] See https://camel-idees.fr/lassociation/le-projet/.

Fig. 8.27 Towed traction for transporting fodder in Rajasthan, India

association of agriculture and livestock. Although they are animals of the great desert areas, large camelids nevertheless participate in oasis or rainfed agriculture on the margins of arid regions. While tradition retains the use of camels to pull the plow (e.g., in Ethiopia), sometimes paired with a mule as in Morocco, to operate the noria and oil mills, or simply to draw water from deep wells as in the Sahel, other more modernized contributions are observed here and there to weed, harrow, lay firebreaks, etc. (Box 8.1).

It is generally considered that the dromedary can plow a hectare in 3 days at a rate of 7 hours per day, to a depth of 16–20 cm. Comparisons with other working species show that the dromedary provides the same results as the horse in terms of power and working time. Furthermore, although the pulse and respiratory rate increase rapidly during exercise, the animal's recovery is rapid, and 2-hour rest after a 4–5 hours' continuous activity cycle is sufficient to restore physiological functions. The interest of large camelids in work functions focuses primarily on the low maintenance cost compared to that of a pair of oxen, especially as their longevity (which varies from 6 to 20 years of productive life) represents a comparative advantage. As with other strength-based activities, training begins at the age of 3. Harnesses can then be placed, either by leaning on the hump (Sudan) or by using the strength of the neck and shoulders (North Africa).

8.5 Camel Transport

If the large camelids are dedicated, in many countries, to the transport of goods and people, it is good to remember that they can also be transported. For a long time, these large animals were only transported on foot. However, the use of trucks, boats, or planes has become commonplace, particularly in the context of the export market for live animals, which are most often destined for slaughter. Motorized transport is also used for racing animals, to access local livestock markets, or more generally for trade (exchanges between farmers). Given the size of the animal, on the road the truck is the most common, but more unusual modes of transport (pickup, car, motorcycle) are possible.

Transport is a major source of stress overall. Numerous studies have shown the impact of the duration and conditions of transport on stress, prompting organizations to consider the welfare of large camelids.[2]

Box 8.1 Dromedary Service Functions (Fig. 8.28)

Indeed, means of transport are often poorly adapted for large camelids, and loading is regularly carried out by less than careful handling (El-Khasmi et al. 2015). The essential knowledge to have to transport an animal according to the recommendations of the Office International des Epizooties (OIE) are the following:

- Space and ventilation requirements as well as feed and water requirements during transport
- Respectful methods of loading and unloading animals
- Understanding animal behavior; signs of illness, stress or distress, pain, and fatigue; and how to avoid them
- Assessment of fitness to travel
- Cleaning and disinfection procedures for means of transport
- Techniques for handling and restraining animals during the different phases of transport (assembly, loading, unloading)
- Methods of managing adverse conditions (weather incidents, road accidents, emergencies)
- The needs according to the physiological status of the animals (young, culled animals, pregnant or lactating females, etc.)

In order to ensure the best possible transport conditions, it is necessary to

[2] Animal's angels, The welfare of dromedary camels during road transport in the middle east, https://animals-angels.de.

Fig. 8.28 (**a**) Camel with a plow in Morocco (paired with a mule). (**b**) Dromedary plowing in Ethiopia. (**c**) Water extraction in Chad. (**d**) Water extraction in Niger. (**e**) Sesame oil mill in Sudan. (**f**) Establishing a firebreak by harrowing in Niger

– Prepare the animals before transport (e.g., training to enter an enclosed area, such as a trailer).
– Evaluate the time of transport and climatic conditions to prepare the quantities of fodder and water accordingly.

Fig. 8.29 Transport of large camelids. (**a**) Crane loading in Turkmenistan. (**b**) Boarding by hoist in Saudi Arabia. (**c**) Stress on boarding in Saudi Arabia. (**d**) Crane landing of a blindfolded male camel in Saudi Arabia. (**e**) Manual boarding on an inclined plane in Turkmenistan. (**f**) Boarding camels on an insufficient surface in Morocco. (**g**) Transport in a pickup truck in Dubai

- Prepare transport vehicles and adapt them to the types of animals, particularly in terms of space.
- Check, during the journey, the sanitary or stress state of the animals (trembling, diarrhea, vocal complaints).
- Provide rest areas during the journey if it exceeds a certain duration.

One of the most delicate points is often the loading procedure. The area should be well prepared and quiet, which is rarely the case. Several techniques are used which are more or less coercive depending on the degree of preparation of the animals, whether they are used to entering a method of conveyance or not (Fig. 8.29), either by force by lifting the animal, by pushing it on an incline, or even by using a crane. Procedures that may cause pain or increased stress, such as the use of a whip or stick or manual restraint by pulling on the ears and lips, should be avoided. In some cases, blindfolds are used to reduce stress, as is the practice with wildlife.

References

Algruin K, Konuspayeva G (2015) Making laben with camel milk. In: Konuspayeva G (ed) Proc. of 4th Conference of ISOCARD, "Silk Road Camel: The Camelids, main stakes for sustainable development", June 8–12, 2015, Almaty, Kazakhstan, Special Issue of Scientific and Practical Journal Veterinariya #2 (42) 2015, poster session, 393

Berhe T, Seifu E, Kurtu MY (2013) Physicochemical properties of butter made from camel milk. Int Dairy J 31:51–54

El-Khasmi M, Chakir Y, Bergaâ R, Barka K, Lektib I, El-Abbadi N, Belhouari A, Faye B (2015) Impact of transport distance on stress biomarkers levels in dromedary camel (Camelus dromedarius). Emir J Food Agric 27(6):507–512

Faye B, Abdelhadi O, Raiymbek G, Kadim I, Hocquette JF (2013) La production de viande de chameau: état des connaissances, situation actuelle et perspectives. INRA Prod Anim 26(3): 247–258

Hasi S, Amu G, Zhang W (2020) Camel hair structure, properties, and commercial products. In: AlHaj O, Faye B, Agrawal RD (eds) Handbook of research on health and environmental benefits of camel products. IGI Global, Hershey, pp 328–347

Hermann K, Fisher A (2004) Method for hygienic slaughter of camel. In: Farah Z, Fisher A (eds). Milk and meat from the camel - Handbook on products and processing. Zurich, VDF Hochschuleverlag publ., 89–108

Kadim I, Mahgoub O, Faye B, Farouk M (2013) Camel meat and meat products. CAB International, Oxfordshire & Boston, p 248

Kappeler SR, van den Brink HJ, Rahbek-Nielsen H, Farah Z, Puhan Z, Hansen EB, Johansen E (2006) Characterizations of recombinant camel chymosin reveals superior properties for the coagulation of bovine and camel milk. Biochem Biophys Res Commun 342:647–654. https://doi.org/10.1016/j.bbrc.2006.02.014

Konuspayeva G (2020) Manufacture and challenges of camel milk cheese. In: El-Haj O, Faye B, Agrawal S (eds) Health and environmental benefits of camel products. IGI Global, Hershey, pp 110–122

Konuspayeva G, Faye B (2021) Recent advances in camel milk processing. Animals 11(4):1045. https://doi.org/10.3390/ani11041045

Konuspayeva G, Lemarie E, Faye B, Loiseau G, Montet D (2008) Fatty acid and cholesterol composition of camel's (Camelus bactrianus, Camelus dromedarius and hybrids) milk in Kazakhstan. Dairy Sci Technol 88:327–340. https://doi.org/10.1051/dst:2008005

Konuspayeva G, Camier B, Aleilawi N, Al-Shumeimyri M, Al-Hammad K, Algruin K, Alshammari F, Beaucher E, Faye B (2017) Manufacture of dry- and brine-salted soft camel cheeses for the camel dairy industry. Intl J Dairy Technol 70(1):92–101. https://doi.org/10.1111/1471-0307.12319

Konuspayeva G, Baubekova A, Akhmetsadykova S, Faye B (2023) Traditional dairy fermented products in Central Asia. Int Dairy J 137:105514. https://doi.org/10.1016/j.idairyj.2022.105514

Raiymbek G, Kadim I, Konuspayeva G, Mahgoub O, Serikbayeva A, Faye B (2015) Discriminant amino-acid components of Bactrian (*Camelus bactrianus*) and Dromedary (*Camelus dromedarius*) meat. J Food Compos Anal 41:194–200. https://doi.org/10.1016/j.jfca.2015.02.006

Wernery U, Fischbach S, Johnson B, Jose S (2008) Evaluation of alkaline phosphatase (ALP), γ-glutamyl transferase (GGT) and lactoperoxidase (LPO) activities for their suitability as markers of camel milk heat inactivation. Milchwissenschaft 63(3):265–267

SPRINGER NATURE

GPSR Compliance

The European Union's (EU) General Product Safety Regulation (GPSR) is a set of rules that requires consumer products to be safe and our obligations to ensure this.

If you have any concerns about our products, you can contact us on ProductSafety@springernature.com

In case Publisher is established outside the EU, the EU authorized representative is:

Springer Nature Customer Service Center GmbH
Europaplatz 3
69115 Heidelberg, Germany